RIVET LAD

MORE BATTLES WITH OLD STEAM BOILERS

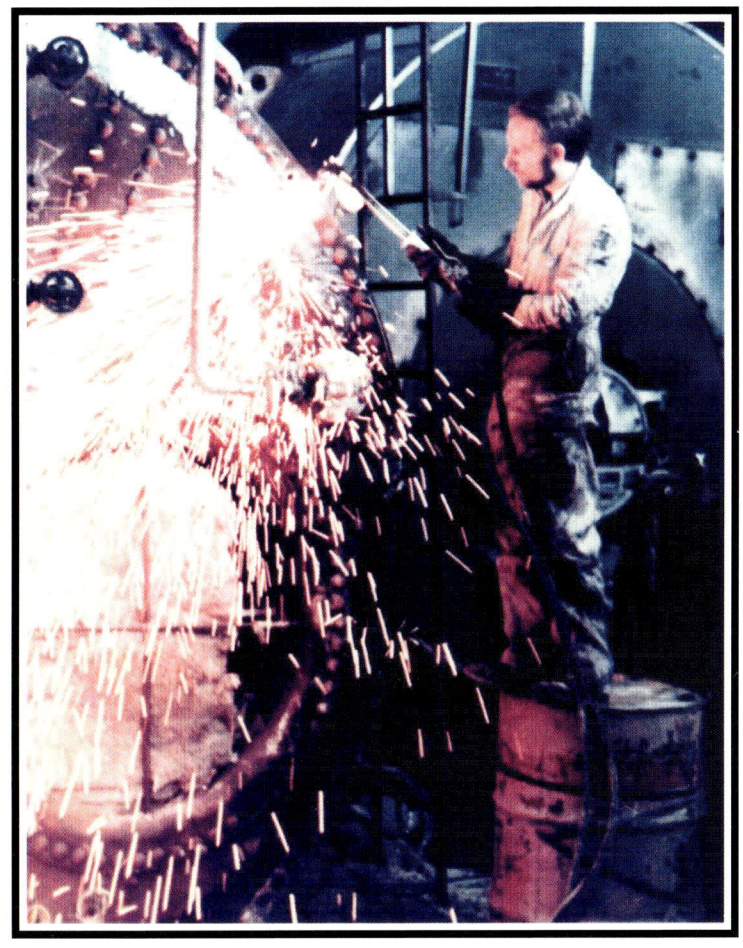

Alan McEwen

****

**WORLD FROM ROUGH STONES HOUSE,
FARLING TOP, COWLING, NORTH YORKSHIRE,
BD22 0NW**

SLEDGEHAMMER ENGINEERING PRESS LTD

www.sledgehammerengineeringpress.co.uk

DEDICATED TO CHRISTINE

MY WIFE AND BEST FRIEND, WHO WITH TYPICAL GOOD CHEER, NOT ONLY TRANSLATED THE HIEROGLYPHICS OF MY MANUSCRIPT, BUT ALSO TYPED AND CARRIED OUT THE FULL DESIGN OF THE BOOK, COVER TO COVER. WITHOUT HER, SIMPLY THERE WOULD NOT BE A BOOK.

ACKNOWLEDGEMENTS:
Richard Bond
Joanne Swarbrick
Geoff Hayes
D.E. Potter
Eric Potter

WITH SPECIAL THANKS TO:
Richard Bond for his cartoons.
Joanne Swarbrick for her painting of Richard Bond's cartoon on the front cover.
www.joanneswarbrickart.co.uk

My sincerest apologies to anyone I have mistakenly excluded.

<u>Front Cover:</u>　　The Author on a job at Edgeley Marsh Pig Farm. The cartoon picture by Richard Bond and the painting thereof by Joanne Swarbrick.

<u>Title Page:</u>　　The Author minus safety glasses and standing upon an empty oil drum, the floor strewn with asbestos lagging, gas-axing rivets on the front-end plate of a steam accumulator converted from a Lancashire boiler. The picture was taken around 1970.

Unless otherwise stated all photographs belong to the Sledgehammer Archive and are Copyright.

ABOUT THE AUTHOR

Author Alan McEwen busy writing.

Alan, foremost is a Boilermaker and Engineer. He is also an enthusiastic researcher, writer and photographer of industrial history which has fascinated him from being a youngster. When he was just 14 years old, Alan researched and produced a small booklet titled, **'THE HOPWOOD COLLIERY WAGGONWAY'**. Regrettably it has been out of print for over 50 years.

In 1977, he researched, wrote and published, **'COLLECTING QUACK CURES'**, and in 1998 wrote **'CHRONICLES OF A LANCASTRIAN BOILERMAKER'**, his autobiography relating his passionate story of how, after serving an apprenticeship as a boilermaker, he became Managing Director and Chief Engineer of which eventually became nationally renowned industrial and heritage boiler engineers, H.A. McEWEN (BOILER REPAIRS) LIMITED, the family firm that he had founded in 1968. Alan retired from the company in November 2007, handing over the reins of the firm that was his life for almost four decades, to his son Alasdair. Alan is immensely proud that under Alasdair's direction, the business now 50 years old is still thriving.

In January 2008, Alan commenced researching and writing: **'FRED DIBNAH'S CHIMNEY DROPS'**. The book was published 15th October 2008.

Back in late February 2009, Alan commenced researching and writing **'HISTORIC STEAM BOILER EXPLOSIONS'**: the subject having fascinated him for a great many years. The book was published August 2009. Alan has also written numerous articles for magazines and books about British Industrial Heritage; he has been a member of the Newcomen Society for almost four decades and the Northern Mill Engine Society for many years.

September 2017 saw the publication of Alan's Boiler Making novel **RIVET LAD – Lusty Tales of Boiler Making in the Lancashire Mill Towns of the 1960s**.

Alan is currently in the throes of researching and writing another book to be titled**: 'THE STEAM CRANE MAKERS OF YORKSHIRE'**. To be published by Sledgehammer Engineering Press Limited.

Also by Sledgehammer Engineering Press Limited:
Chronicles of a Lancastrian Boilermaker, 1998. ISBN: 978-0-9532725-0-8
Fred Dibnah's Chimney Drops. 2008. ISBN: 978-0-9532725-1-8
Historic Steam Boiler Explosions. 2009. ISBN: 978-0-9532725-2-5
Jaggermen's Bridges on Packhorse Trails
    by Christine McEwen, 2010. ISBN: 978-0-9532725-3-2
Rivet Lad, 2017. ISBN: 978-0-9532725-4-9

**Rivet Lad – More Battle with Old Steam
    Boilers. 2018.** ISBN: 978-0-9532725-5-6

The right of Alan McEwen to be identified as the Author of this Work has been asserted by him in accordance with the Copyright, Designs and Patents Act 1988.

All characters in this book, excepting the Author, are fictitious and any resemblance to real persons, living or dead is purely coincidental.

Book layout and design by Christine McEwen.

All rights reserved. No part of this publication may be reproduced, stored in a retrieval system or transmitted in any form or by any means (electronic, digital, mechanical, photocopying, recording or otherwise) without prior permission of the publisher.

Published by Sledgehammer Engineering Press Limited, World From Rough Stones House, Farling Top, Cowling, North Yorkshire, BD22 0NW, England, Telephone: 01535 637153   Email: lankyboilermaker@btconnect.com
www.sledgehammerengineeringpress.co.uk
All distribution enquiries should be addressed to the publisher.

Printed by Amadeus Press, Ezra House, West 26 Business Park, Cleckheaton, West Yorkshire, BD19 4TQ
Telephone: 01274 863210: Fax: 01274 863211: Email: info@ amadeuspress.co.uk   Website: amadeuspress.co.uk

RIVET LAD - More Battles with Old Steam Boilers

CONTENTS

	PAGE NUMBERS
Title Page	i
Captions	ii
Acknowledgements	ii
The Author	iii - iv
Contents	v
Glossary	vi
1 My Battles with an Old Boiler and an Angry Sow at Dandelion Colliery Piggeries.	1-22
2 Edgeley Marsh Piggeries Job.	23-34
3 My Return to Dandelion Colliery Piggeries Job.	35-45
4 World's End Piggeries Job.	46-59
5 The Black Bird Chocolate Works Job.	60-73
6 The Old Golden Pits Piggery Job.	74-89
7 Bessie Roggerham's Bluebell Nook Piggery Job.	90-108
8 An Old Boiler Maker's Scrapbook of Old Boilers.	109-119
9 Other Quality Books From Sledgehammer Engineering Press Ltd.	120
10 H.A. McEwen (Boiler Repairs) Ltd. 50 Years Celebration advert.	121
11 Coming Soon!	122

GLOSSARY

Ash-wheeler.	*Boiler House Labourer.*
Blackjacks.	*Local name for Cockroaches.*
Boilt t'ham flourcake.	*Lancashire dialect for boiled ham muffin.*
Boyer 60/80.	*Consolidated Pneumatic Co. built pneumatic riveting hammers.*
Connies.	*Feedwater Economisers. (Green's or Goodbrand's)*
Clout.	*A piece of rag, cleaning mop.*
Caulking.	*Upsetting the lap-edge of a boilerplate to form a seal, either by hand tooling or by air-powered caulking machine.*
Crocodile-gob.	*A Vee-mouthed steel rod used for driving tubes out of tube plates.*
Engine-tenter.	*Mill steam engine operator.*
Fire-beater.	*Lancashire term for boiler fireman or stoker.*
Ginnel.	*An alleyway or back entry.*
Gobbing.	*The cutting by oxy-propane burning torch of wide mouth-shaped sections of smoke tubes to facilitate ease of withdrawal from the tube plates.*
Gas Axe.	*Oxy-propane burning torch.*
Jack-bit.	*Meal break or food.*
Kier.	*A riveted iron or steel pressure vessel of spherical or horizontal, cylindrical, dome-ended construction for boiling paper pulp, rags etc.*
Maul.	*A heavy, long-shafted hammer.*
Mill-wrightery.	*The power transition shafting and gearing within a mill.*
Magic Fluid.	*Steam.*
Mud-hole.	*Aperture in boiler for inspection and cleaning.*
Man-hole.	*Most boilers have one or more for allowing access into the interior.*
Mule's Ear.	*Boilermaker's vernacular for a steel lifting lug.*
Mystic.	*Coolant oil for metal cutting.*
Snap.	*Meal break or food.*
Snap Bothy.	*Eating place in a boiler shop.*
Sprag.	*Heavy steel bar used for jamming or rounding up steel cylinders, boiler shells etc.*
Swale's Swill.	*Nickname for a Lancashire beer.*
Swan's neck.	*Cast-iron communication pipe from blowdown block to blowdown valve.*

Chapter 1

MY BATTLES WITH AN OLD BOILER AND AN ANGRY SOW AT DANDELION COLLIERY PIGGERIES

PHOENIX BOILER MAKERS

Upon completing my Boiler-making apprenticeship with Carrot Crampthorn's colourful and trusty squad, I remained grafting alongside his merry band of Boiler Makers until my departure from Phoenix Boiler Makers in 1968, when I established my own small boiler repair firm: H.A. McEwen (Boiler Repairs).

Indeed, I fondly recollect the day I kicked off to establish myself, which was the beautiful, gloriously sunny morning of the 4th of August. After I had acquired a considerable amount of specialised Boiler-making tools – hand-hammers, sledge-hammers, pinch-bars, podger-spanners, tube expanders and mandrels, and importantly a set of oxygen and propane burning gear - the source mainly being retired Boiler Makers, the next important acquisition was a suitable, yet extremely cheap van, required for the transportation of myself and my heavy steel toolboxes.

Thus, for the princely sum of £8, I managed to purchase from Gordon, the local milkman, a dark green, 1954, Austin Devon estate, registration number KPM 500. My milkman mate evidently had previously used the vehicle for delivering morning pints of milk. The dark green estate was blessed with a truly diabolical steering column gear lever – (3 forward and reverse). On changing gear, the cacophony of gear-crunching escaping from the interior of the exceptionally worn gearbox, made me wince.

The old Austin Devon, despite its woebegone appearance, was a most reliable starter, either via the battery, or by the starting handle! Upon her engine being thoroughly warmed through, she would happily motor along all day at her maximum speed of 38 m.p.h. She had however, a terrible thirst for petrol, averaging about 12 miles per gallon, and her rather tired old engine consumed oil by the gallon. The van's noisy exhaust system generated dense clouds of choking white smoke.

Nevertheless, the old lass proved to be a damned good, reliable, workhorse whilst carrying heavy boxes containing my Boiler Maker's tools, bundles of smoke tubes and heavy slabs of boiler plate to sites all over Lancashire and the West Riding of Yorkshire. The Devon was quite a curious-looking vehicle too, for typically, on hot summer days, when seen from a distance - due to her old brass radiator being prone to over-heating which produced clouds of billowing steam – the vehicle could be mistaken for a steam-powered car!

I clearly recall one hot and sunny, early September afternoon, the Devon estate chugging up the vertiginous climb from Littleborough over Blackstone Edge on the A58, the mountainous, trans-Pennine road leading into the West Riding of Yorkshire, where I was to do battle with a troublesome steam leak on a Lancashire boiler powering a Sowerby Bridge woollen mill. The Devon's big brass radiator was running extremely hot, so I had removed the cap.

With both hands holding onto the violently bucking steering wheel, peering through the windscreen I observed, spewing out of the cap-less radiator enormous gouts of white steam, puffing out like a hard-at-work saddle tank steam locomotive. The steam enveloped the Devon's cab through the open windows obscuring the way ahead and making me violently cough. Halcyon days!

Among the first jobs I typically attended to, ranged from renewing a leaking smoke tube in a Foster, Yates & Thom Economic horizontal, dry-back boiler or, 'buttering': building up by electric-arc welding, wasted McNeil mud-lid compensation plates on a Cochran Vertical boiler; hand-caulking small, yet annoying leaks on riveted joints on Lancashire and Cornish boilers; changing over leaking fusible plugs and carrying out diverse minor welding repairs to steam pipe work and related machinery.

BOARSHAW BOILERWORKS

In 1969, I lustily converted an ancient timber and brick-constructed double-fronted garage situated on busy Boarshaw Road, Middleton, into my very own Boilershop, whilst handing over the weekly rent of £2. 10 shillings to a local butcher who owned the property.

I subsequently installed electricity and a telephone. This telephone was my very first means of communication. The phone number was MID 2008. For water required for the hydraulic testing of boilers – (I couldn't afford the outrageous figure quoted by the local Water Board) – thus, from an old electric-powered Worthington-Simpson boiler feed water pump and several lengths of 2 inch iron piping scrounged from a local cotton mill, I knocked up a Heath Robinson water pumping system. Upon completion, my pump delivered all the water I needed from the Wince Brook – (locally called Jelly Brook, after a local worthy whose surname was indeed Jelly!) - that flowed at the rear of my Boilershop. I stored the pumped water – which occasionally reeked of waste dye work's effluent – in a large, riveted galvanised iron tank I had installed up in the roof space. When full, the tank induced grunts and groans from the worm-eaten rafters due to the considerable weight of water.

Between a few minor boiler repair jobs, which I attended to in the local cotton mills and factories which granted me a sparse living and kept me sane, I supplemented my meagre earnings by trying my hand at chassis and body repairs on cars, vans and the odd lorry. I was proficient at oxy-acetylene welding, wasn't overly concerned about lying for hours on end beneath filthy, oily old bangers; wasn't frightened of getting burnt from molten rubber undersealing compound, which I frequently did; and wasn't too afraid when the oxy-acetylene gas welding bottles due to a nozzle blow-back would suddenly erupt in gigantic flames!

For months I practiced hard and word soon got around, resulting in me being sought after by owners of vehicles requiring panel beating, or for the welding of large steel patches over massive holes in the bodywork of Bedford coal lorries, Commer vans and such like. Several profitable regular jobs came from a local asphalt contractor who operated a fleet of lorry-hauled, 8-ton, coke-fired tar-melters.

Asphalt George, as I called him, would usually ring me up at around three o'clock in the afternoon to inquire if I could, on an urgent, drop-everything-else basis, replace the burnt-out, trough-shaped, pan-bottom of one of his firm's tar-melters and, have it completed for mid-night! George would expect me to give him an unequivocal, cast-iron promise that the tar-melter would be finished on time, thus enabling them to collect it at six o'clock the very next morning. Rising to the challenge, and perhaps more importantly, badly needing to earn the brass, I always answered

Asphalt George in the affirmative, and by about tea-time, a massive, reeking tar-melter would be towed by a 14-ton Ford lorry to be parked on the broken flagstones outside my Boilershop.

The tar-melters were so huge that I couldn't manoeuvre them inside the Boiler Shop. I was forced therefore, to attend to the burning out of the badly burnt pan-bottom plate-work and the fabrication and the welding-in of the new ⅜ inch thick plates, out on the Queen's Highway in the fresh night air. Thus, armed with two or three, 240 volt electric hand lamps to lighten up the gloom, with great gusto, I would get stuck in with all the necessary violence of an enthusiastic, desperate-to-earn-money young Boiler Maker. The sound of my trusty, short-shafted, 7 lb sledgehammer went **'boom, boom, boom'**, as I hammered the steel plates, the terrific noise reverberating around the tightly packed surrounding streets of terraced dwellings, unfortunately, causing much distress to my neighbours.

During the snowy, cold winter of 1969-1970, I repaired about twelve tar-melters which made me quite a few bob. It was indeed satisfying, profitable work. Here I was, a young self-employed Boiler Maker, working in the craft I loved and in my own little Boarshaw Boilerworks; the upsetting of the neighbours notwithstanding!

However, things were soon to change for the worse. One morning, a rather sad-looking, sallow-faced, officious individual, attired in a grey suit and wearing a grey trilby, called to see me at the Boilerworks. The grey little man, quite obviously pleased himself upon handing me a 'Noise Abatement Notice' from the Council. This document forbade me carrying on with my nocturnal, roadside engineering activities, full stop! The Council's man mentioned Strangeways Prison! Hell's Teeth!

Upon me reading the official wording of the document, it induced me to think again about my lorry chassis welding and tar-melter re-bottoming nocturnal activities. For despite the regular good money I had earned, it wasn't 'real Boilermaking', and after all, it wasn't what I had hoped to achieve after enduring six, long, hard years as an apprentice at Phoenix Boilermakers. I also didn't fancy a stint in Her Majesty's Strangeways Jail!

Clearly, what I did miss, was that wondrous feeling of pride that I had experienced when a broken-down Lancashire boiler or a locomotive boiler was subsequently returned to steam as a result of me carrying out some highly-skilled Boilermaking work. Indeed, I was a time-served Boiler Maker and proud of my skills!

For several weeks thereafter, my fledgling business suffered as I was forced to turn down many profitable, yet noisy repair jobs on lorries and tar-melters. This resulted in me feeling a wee bit debagged.

ARDPHALT ASPHALT & CO. LTD.
THE ASPHALT FLOORING & ROOFING SPECIALISTS

Over 25 year's experience

INDUSTRIAL, COMMERCIAL & DOMESTIC WORK OUR SPECIALITY

**Bradshaw Trading Estate, Greengate, Middleton, Lancs.
Tel. 061-643 7040**

Competitive prices, a work guaranteed
Distance no object, all areas covered
Free estimates given

A 1960s advertisement for Ardphalt Asphalt & Co. Ltd.

However, Lady Luck would soon smile upon me. For one morning in October 1969, I received a phone call from the manager of a Manchester-based steam-engineering company. In his rasping Mancunian brogue - for the manager must have smoked a 100 Capstan Full Strength cigarettes per day – informed me that he had noted down my name and phone number which was proudly displayed on bright orange coloured sheets of paper glued onto the Austin Devon's rear side windows, while the van was parked at the entrance of Manchester Victoria Railway Station.

My new found, dry-throated friend asked me if I wanted to attend to a profitable boiler repair job at a pig farm. He explained his own firm were far too busy. I later learned this was an excuse. His requirement involved me burning out the heavily-corroded and leaking uptake tube in a John Bown, vertical cross-tube boiler belonging to a pig farmer at Dandelion Colliery Piggeries in Claybank. I replied in the affirmative, informing Smokey Joe that I would attend to the job that very day.

I was indeed well experienced in carrying out all manner of jobs on vertical cross-tube boilers, which were often much abused. I previously recall during my time with Carrot Crampthorn's Squad, carrying out a multitude of repairs; such as the replacement of fire boxes, the renewal of uptake tubes, the blanking off of leaking, corroded cross-tubes, building up metal wastage around the McNeil mud-lid and man-lid compensation plates; replacement of corroded rivets, and the changing over of leaking fusible plugs.

Thus, this job at Dandelion Colliery Piggeries would be my introduction into the world of 'the Piggers' and their clapped-out boilers which would, I hoped would eventually lead to some exciting and immensely profitable work.

Despite my youth, because my former calling with Phoenix Boiler Makers had taken me into factory sites all over Great Britain and Ireland, I was therefore, very well acquainted with meeting diverse types of people while carrying out my daily craft. For I had indeed, met virtually every race and creed; many gregarious, colourful characters; ever joyful types and also the opposites; dull, mean-spirited, down-right sad and rather miserable folk: both male and female. However, there is one group of hard-working Lancastrians indelibly printed within my mind: 'The Piggers'.

Pork fatteners, pig-feeders, pig-rearers, swine herds – or simply, pig farmers. Call them what name you like, but in 'Silly Country' – as the east Manchester suburb of Claybank was locally known, farmers engaged in the production of British prime pork, were known in the local vernacular as 'The Piggers'.

Upon gratefully receiving the much-valued tip-off from the rasping – voiced Mancunian on the telephone, I got stuck in with gusto loading my heavy, ex-British army steel ammunition boxes serving as tool boxes which held a vast assortment of caulking chisels, tube expanders, light and heavy hand hammers, pinch-bars and a couple of 7-lb, short-shafted sledgehammers into the back of the Austin Devon van.

One of the tool boxes contained an assortment of Walker's asbestos boiler man-hole joint rings, and several tins of manganesite jointing paste. On the righthand side of the stack of tool boxes, I heaved into position a large and heavy oxygen cylinder; on the left I placed a tall red-painted propane cylinder. The rolled-up rubber burning pipes, oxygen and propane gauges and my trusty 'gas-axe' – the burning torch – was placed on top of the tool boxes. Finally, with great difficulty I rammed on top of the load, which almost scraped the van's roof a 6-foot length of 9-inch diameter by ½ inch wall, seamless carbon steel tube. This material was a replacement uptake tube for the vertical cross-tube boiler.

Reaching down at the front of the Devon's bulbous radiator housing, and grasping the rusty iron starting handle – for the ancient battery power was low – I cranked up the tired, old oil-burning Austin Devon's engine, which coughed and spluttered, then roared into life.

With a clunk I engaged first gear and the van lurched forward along Boarshaw Road heading towards Middleton town centre. Within less than an hour I found myself driving through Failsworth en-route to Claybank.

John Bown & Co.'s, Dukinfield Iron Works. Originally Bown & Co. built Lancashire and Cornish boilers. They also built Vertical Cross-tube Boilers.

A JOHN BOWN VERTICAL CROSS-TUBE BOILER AT DANDELION COLLIERY PIGGERIES

I fondly recall that the day had opened out into a very pleasant, warm, golden October afternoon as I slowly drove the battered Devon van through Claybank, a working-class district of streets of Accrington brick terraced houses and Victorian-built cotton mills and factories. My destination was the Jackson Brothers' boiler-shed at Dandelion Colliery Piggeries. Turning off the main road, I then headed downhill towards where the canal and the railway line linking Manchester with Yorkshire - on the other side of the Pennines - competed for space in the tight confines of the valley bottom.

Arriving at the bottom of a cinder track, on seeing a much-weathered sign: **Jackson Brothers, Pig Fatteners**, I drove through a broken iron gate and into a cobblestoned yard which hosted several semi-derelict buildings. After parking the Devon and looking about, I could see that the Piggery consisted of around half an acre and was sited alongside a rubbish-filled, disused canal. The pigsties themselves were fenced with a motley collection of old enamel advertising signs – **'Turf Cigarettes, Robin Starch and Zebra Polish'** - were three such signs out of many.

Other fencing materials consisted of old railway sleepers, sheets of wriggly tin, and doors and roofs of clapped-out old lorries and vans. I noticed also, the rusting remains of a large, portable steam engine which sported a huge Ramsbottom safety valve amongst a clump of large elderberry bushes and brambles. The whole place resembled a scrap metal junk yard! And an exceedingly strong reek of pig-ordure pervaded the whole premises.

Suddenly I heard a shouted greeting. ***"About time! Tha'll be th' new boiler man. Get thee gear unloaded quickly, we're desperate for steam!"***

Quickly turning around in the direction of the angry voice, I spied an elderly man emerging out from one of the dilapidated buildings. I walked towards him, introduced myself whilst offering my hand. The old man declined to shake my hand. He introduced himself as Hiram Jackson. Mr. Jackson looked around 70 years of age, and was a tall, walking-stick thin chap, sporting a curious-looking long nose, and cornflower blue eyes that peeped out from a yellowish face of parchment-like skin.

Poking out from beneath his tweedy cap were long, wispy strands of dirty, white hair. He was strangely attired in an ancient-looking, pig-muck stained dark blue striped jacket and trousers. His footwear was a pair of pig-muck plastered turned-down wellington boots.

He spoke with a rasping smoker's voice; a virtual continuous stream of expletives poured rapidly from between thin, cruel lips; his teeth, a nicotine yellow-brown, the result of what I guessed was, fifty-odd years of smoking strong cigarettes. I noticed with disgust, the elderly Pigger, after barking out a sentence would then cough and eject from his mouth a vile stream of brownish phlegm. He fixedly stared at me in a mesmerising manner as though he was demented. Even though I had only just met Mr. Jackson, I clearly considered him to be a most unpleasant breed of chap.

Turning around towards the buildings, Jackson shouted an instruction, **"Marcus, get thee backside oot o' yon chair an' meet our new boiler man."**

A heavily built man in his mid-forties quickly scurried out of a nearby building and extended his hand towards me whilst greeting me with a huge smile, **"Hiya cock, welcome to Jackson Brothers' pig fattening farm"**, he breezily exhorted. **"I'm Hiram's son, Marcus."**

Marcus had long muscular arms and unusually short, thick legs giving him a bulky-squat appearance. He was dressed in a much-patched pair of light blue bib and brace overalls and an old ragged-collared check patterned shirt. Upon his head, at a jaunty angle, he sported a cap of light blue denim material. He did however, appear to be a very pleasant chap, and totally different to his horrible father.

THE SWILL HOUSE

Hiram, darkly scowling and pointing down the yard, then instructed me to park my Devon van alongside the Swill House. Built of rotting brick, about 12 feet square and with a roof of corrugated iron, the Swill House was sited at the bottom of the filth-strewn yard. I noticed a rusty iron chimney poking skywards through the roof, which no doubt, I considered, was mounted upon the John Bown Vertical Cross-tube boiler I had come to repair.

A typical design of a Vertical Cross-tube Steam Boiler.

Marcus Jackson led me inside the building. Looking about me, I noticed there was only a small, two-foot square window, the filthy panes of glass were blanketed in a smattering of dead flies and spiders. The single electric light bulb hanging from a wooden beam was the only source of illumination. Standing at the left-hand side of the door was a 13 feet tall by 5 feet diameter, Bown Vertical Cross-tube boiler.

Upon reading the tarnished brass maker's plate informed that the boiler had been built in 1926 by John Bown & Co. Dukinfield. At the side of the boiler, mounted upon brickwork was a riveted, galvanised iron trough used as a feed water tank for replenishing the boiler's water. An ½ inch pipe dropped down from the feed water valve on the boiler into the tank. I could see there was a live-steam water injector of ancient origin among a cluster of rusted iron pipe work.

Lined up against the rear wall were five huge, old wooden beer casks which Marcus explained were used for boiling the pig swill. A 2 inch iron pipe ran from the crown valve on the domed roof of the boiler around the inside of the room and was fastened to the wall about 3 feet above the casks. Interspersed along this pipe were several steam valves from which 1 inch pipe down-comers dropped into the depths of each cask for cooking the pig swill.

Stacked high all around the inside walls of the Swill House were an amazing collection of early wireless sets, old beer crates and flimsy wooden soap boxes. Jammed inside these boxes were bundles of yellowing newspapers. Looking about me in the gloom, as I moved around the boiler jotting down various dimensions in my notebook, I was surprised to see a hen sitting upon eggs on the top of an old apple box, the creature making a series of clucking noises, its head moving to and fro, its eyes staring at me - the Swill House invader.

The whole ram-shackle building gave off the most offensive odour; a sweet and sour concoction of stale pig swill and rotting food. The floor was inches deep in a squelching, glutinous, slippery mess of ground-in swill and pig muck. Scattered around the door out in the yard was a couple of dozen open-topped metal dust bins that were brim-full of a stinking, rotting mess of congealed 'school dinners', cotton mill canteen food waste, and the contents of waste food slop buckets that the Jacksons had brought in from their neighbours' homes on the nearby council estate. The whole of the Piggery stank Heaven High, and despite me being well used to revolting working conditions, the Jackson's Dandelion Colliery Piggeries, really was a most horrible place to work.

A pair of Ruston & Hornsby Ltd Vertical 'Thermax' Boilers.

MARSHALL VERTICAL BOILERS

Built with a margin of strength and safety, which ensures economical service and long life with freedom from breakdown

MADE IN SIZES TO PROVIDE EVAPORATIONS FROM 225 lbs. to 1,745 lbs.

Deliveries from Stock.

A Marshall 'Trent' VCT Boiler.

REPLACEMENT OF THE UPTAKE TUBE

With difficulty, due to the dangerous slippery surface of the yard immediately in front of the Swill House, I quickly unloaded my oxy-propane burning gear, electric welding set, cables and necessary hand tools from the Austin Devon. Marcus wired up the welding set into an ancient-looking electricity supply panel mounted on the wall. After I connected up the oxy-propane 'gas-axe', I then got started on the dismantling of the boiler's iron chimney which was secured by four ¾ inch Whitworth bolts on a flange rising from the centre of the dished end plate. From this position the chimney jutted skywards through the wriggly tin roof, its overall height being around 8 feet.

OWD WALTER JACKSON

I was just about to light up the 'gas-axe', when a sudden movement among the surrounding floor-to-ceiling stack of old wirelesses and boxes, made me stop and look behind me. Emerging from the semi-gloom of the Swill House interior, was what appeared to be an extremely diminutive, elderly man whose head and face was hidden behind a piece of hessian sack cloth. As the old man shuffled towards me I noticed that in his hand he was carrying a large mug of steaming hot liquid. The man then spoke, **"A mug o' warming tea for thee, lad. Our Hiram told me to mek thee a brew."** He placed the mug atop the feed water tank and extended his hand towards me. We shook hands. **"They call me Owd Walter, I'm Hiram's brother and partner. Welcome son, you'll be th' new Boilerman?"**

Looking down on the short, old man whom I could see was bizarrely dressed in a mixture of ragged, grubby male and female clothing, I noticed also with curiosity that when he spoke he would keep pulling the sacking around his head and face. He then shuffled out of the Swill House.

Using the burning torch, I quickly sliced through the chimney bolts and then after going outside into the yard, I climbed a wooden ladder propped up against the wall of the Swill House and clambered onto the wriggly tin roof. My plan was to wrap my arms around the now unsecured, thin iron chimney and bodily lift it up through the hole in the roof and then to lay it safely down. However, my plan quickly became a disaster!

Grabbing the rusty steel chimney in a bear hug, I joggled it from side to side which quickly released it. Taking a huge breath of air, I then lifted it upwards. To my surprise the rusty old chimney jolted sky wards faster than I had planned, catching me unawares. It quickly toppled over causing me to lose my balance on the much weathered, greasy surface of the tin roof. To my horror, I felt myself flying through thin air, off the roof. I was terrified that the iron chimney would fall onto me, but fortunately it didn't happen. Instead, I felt myself dropping towards the ground. Splodge, splat! I came down to earth fortunately without injury, but to find myself sprawled upon a thick carpet of reeking, freshly deposited pig muck. I was plastered head to toe in the reeking stuff.

"Eeh, you silly young bugger, you've given Winnie, me best sow a reet bad fright. I'll hev to knock some brass off your bill." Feeling shocked and close to vomiting due to the diabolical reek coming from the sludgy mess of pig muck covering my boilersuit, my hands, face and hair, and with much slipping and sliding I got back onto my feet. To my abject horror I realised I had landed inside a pig-sty. Standing behind the pig-sty fence, I saw both Hiram and son Marcus were grinning and cracking jokes about my plight. **"I trust you've not damaged me boiler chimney,"** said Hiram in his cruel, rasping voice. I then heard another voice, **"Ye cruel pair o' miserable sods. Leave off tormenting th' young Boilerman. Get thee sell' out of yon sty, lad. I'll run thee some hot watter for a bath."**

Rubbing the evil, smelling pig muck from around my eyes with a piece of rag I found in my pocket, I then noticed the kindly voice belonged to the short, friendly old chap, Owd Walter.

Despite being covered in filth, and shaking somewhat, I was pleased that I had not broken any bones. I wasn't even bruised. However, my pride was severely injured! Whilst looking for the gate in the four feet high fence surrounding the pig-sty, suddenly I heard deep grunting sounds emanating from the sty's gloomy interior. With mounting terror, I found myself being confronted with a massive, angry sow, that rapidly charged over to me. I was terrified, for I had some knowledge of what angry pigs can do to the human body.

A pig farmer had once upset me whilst relating the tale of how a huge, angry sow upon being disturbed had attacked a builder, who in the throes of carrying out a building job had innocently entered the pig-sty. The extremely nasty tempered sow had charged at the unfortunate builder and sinking her teeth into one of his buttocks had bitten off a large piece of flesh. Ouch!

As I backed up against the fence, the huge pig, grunting and snarling came towards me. **"Gerr off Winnie, you brute,"** followed by the sound of a timber batten being whacked across the pig's massive back. It was Marcus clutching a length of timber. **"Coom on lad, get out of theer,"** he said while opening the gate. Shaking violently with terror, Marcus then led me back inside the Swill House.

Waiting inside was Owd Walter who handed me a huge mug of strong, sugary tea. **"Get that brew inside thee lad. Tha con get bathed in't back room. I've drawn thee a cask full of hot watter."** Quaffing slugs of hot tea, I followed the tiny, elderly man through a door and entered a small, rear room which I discovered was moderately clean and tidy. Over in a corner was a cut down wooden cask, full of hot soapy water. **"Thee hev a good scrub, I'll give thee some privacy,"** said Owd Walter, whilst pulling the sacking across his head and face.

I must admit, I thoroughly enjoyed bathing in the cask which settled my shattered nerves. Upon clambering out, I noticed set out on a table was a pair of clean, old jeans, a shirt and a pair of woollen socks. Examining the clothes, I could make out they were around my size, and with having little choice but to wear the rather old, but reasonably clean garments, I happily climbed into them.

I was just about to recommence my boiler repair activities, when Owd Walter came back into the Swill House clutching a heaped plate of Jaffa cakes. **"I'm sure tha'll be hungry. Get them orangey biscuits down thee neck,"** he said, then left the room. Marcus suddenly appeared chomping on a handful of Jaffa cakes.

Noticing I had commenced eating, he said, *"Aye, tha con eat as many as you like. We've getten barrels full on 'em. We visit the local Biscuit Manufacturing Works twice a week to collect all their reject biscuits, cake and waste fat which amounts to around 2 tons per load. The stuff meks reet good pig swill. The Jaffa cakes are fresh, they were baked just last neet,"* he said smilingly.

I thought to myself that Marcus was quite a pleasant chap just like his uncle, Owd Walter. I then asked Marcus the reason why his uncle went about with his head and face hidden from view, covered with a piece of smelly old sacking.

"During th' war, Uncle Walter was in a tank regiment. His tank got hit by an enemy shell incinerating all of the crew except him, who, although suffering severe burns to his face and head, managed to climb out of the turret and escape. He is a dreadful sight, and therefore, won't let anyone see his condition, including even mi Dad and me," He said with sadness.

THE UPTAKE JOB

Thus, now well-scrubbed and with my hunger pangs abated, returning to the job in hand, I recommenced the boiler repair operations. Using a mighty 1¼ inch Whitworth podger-spanner, together with a 4 feet length of iron pipe to assist leverage, pulling down hard on each of the nuts securing the boiler man-lid, they were sufficiently slackened off and quickly removed. After lifting off each of the heavy cast-iron dogs, I rained hammer blows onto the centre of the man-lid which broke the joint.

Standing on a short wooden ladder, with Herculaneum effort, I manoeuvred the heavy man-lid out of the boiler and placed it onto a small bench. Later, I would overhaul, clean and fit a replacement asbestos joint ring. NOTE: in those far off days asbestos was commonly used as gaskets and particularly for lagging boilers and pipe work. Therefore, there were volumes of the deadly material in virtually every boiler house.

Donning my burning goggles and leather welder's gauntlets, I then tackled the job of cutting out the condemned uptake tube at the firebox crown attachment, which overtime, had assumed the appearance of a scaly, brownish-orange hued porcupine due to the large assortment of bolts, nuts and lead washers previously having been fitted to arrest steam leakage from the many deep pits covering the surface: the result of oxygen corrosion.

The torch's high temperature cutting flame caused the thick scale to explode into countless red-hot shards which landed on the bare skin of the nape of my neck. The pain was awesome! I was also troubled by the thick black fumes that entered my nose and throat making me cough and retch. But, alas, this was my chosen profession, boiler making, the work that I loved.

By the time I had managed to cut around the radiused firebox crown plate, thus releasing the uptake tube's bottom connection, the skin on the back of my neck was throbbing from the numerous painful burns. By standing on the ladder leant up against the boiler, I then proceeded to chop out the top connection where the uptake emerged out of the outer domed boiler end plate. The uptake was now loose. However, due to it being cumbersome and heavy, I decided to chop it into several manageable sections, which I then manually lifted out of the boiler and moved out into the yard.

The next stage involved using my antiquated electric grinder to grind out weld preps around the top and bottom connections in readiness for the fitting of the new nine-inch uptake tube. With considerable apprehension and fear I clambered onto the Swill House roof where I had earlier left the new uptake tube and tied a short section of rope around the top of the tube.

With much huffing and puffing due to the heavy weight of the steel tube, I gingerly fed it through the hole in the wriggly tin roof and lowered it down until it entered the hole I had prepared in the boiler's domed outer end plate. Easily entering, I lowered it further until I sensed the bottom edge had entered the likewise prepared hole in the domed firebox crown plate where it then came to rest onto the three small steel pegs I had tack-welded into position to act as temporary supports. I later removed these pegs.

Once achieved, and after I had tack-welded the uptake tube into position, I then proceeded with the welding which I achieved by laying down three consecutive deeply penetrating runs of weld metal to form heavy fillet-welds. Being accessible from my ladder, the boiler end plate weld I soon completed. However, the bottom fillet-weld of the uptake tube to firebox crown attachment was exceedingly difficult. This was due to there being an absolute minimum of room within the steam space. Thus, this three-pass fillet-weld took a couple of hours to complete.

Being a very slim, extremely fit short chap of about 8½ stone, squeezing my body through the man-hole and wrapping myself around the uptake tube wasn't a problem. The difficulty was that I had to position my back up against the inside of the boiler shell with my legs bent, knees up on either side of the uptake. This very awkward position was bad enough, but I was also troubled due to torrents of sweat pouring down my forehead to drip off the end of my nose, making my eyes smart.

With great difficulty I also had to place the welding helmet over my tweedy cap followed by the pulling on of the heavy, thick leather gauntlets. Nonetheless, after much puffing and panting in the extremely hot and smoky confines of the steam space eventually, with extreme effort and skill, I satisfactorily completed the welds.

After taking a short rest and consuming another pint of Owd Walter's hot sugary tea, I completed the boiler repair work by installing a new NABIC fusible plug. This was followed by overhauling the McNeil-type man-lid and the fitting of a new asbestos joint ring. With difficulty and sweating profusely, I carried the heavy man-lid up the ladder and fitted it into the boiler shell. Holding it in position by grasping one of the threaded bolts, I then fitted both cast-iron dogs and the nuts. Both were nipped up tightly. Marcus, after pushing a rubber hose pipe into a 1 inch screwed boss on top of the boiler, then commenced filling the boiler with cold water from the Pig Farm's supply pipeline. Upon being filled, he withdrew the hose and then screwed in the plug.

The last job was to prove that my work was sound, thus, using my hydraulic test pump, I applied a pressure test which proved the welds were tight.

Marcus kindly assisted me to remount the iron chimney, and also to replace the old fire bars made from sections of railway line, forming the grate. He then delivered several wheel barrow loads of dry wood off-cuts and a huge cardboard tub full of the local Biscuit Work's waste butter fat. After draining off the boiler feed water to the correct working level, I then built up the firewood onto the fire bars. Throwing in a blazing diesel-soaked rag, quickly produced a fantastic conflagration.

About an hour later, Owd Walter, Marcus and even miserable old Hiram were all smiles as we sipped our mugs of hot sugary tea, whilst chomping on Jaffa cakes grouped around the John Bown Vertical Cross-tube boiler, which was by now merrily steaming at 30 p.s.i. Marcus, chuckling to himself, kept adding slabs of butter fat onto the blazing wood fuel, making the boiler roar.

Venturing out into the dark night, I could see gigantic gouts of bright yellow flames spewing out of the top of the Swill House chimney to light up the surrounding darkness of Dandelion Colliery Piggery. Because the boiler was apparently uninsured – so typical of the day - the Jackson Brothers would not be receiving a visit from a Boiler Inspector. Boiler insurance was frowned upon by the Piggers! I was however, totally confident in my workmanship.

The Jackson family were delighted with the boiler repair work and paid me the then princely sum of £45.10 shillings. Marcus loaded the Austin Devon with two huge boxes of Jaffa cakes, whilst Hiram informed me that he would broadcast my name to other pig farmers all over Manchester and beyond.

He was true to his word, for subsequently I would carry out numerous boiler repairs at Piggeries, small dairies, tomato growers and sundry horticultural firms all over Manchester, Lancashire and the West Riding of Yorkshire.

Vertical Cross-Tube Boilers were used in Pig Farms, Dairies, Horticulturalists and other industries requiring small amounts of low-pressure steam.

Chapter 2

EDGELEY MARSH PIG FARM JOB.

One freezing cold December evening in 1969, after carrying out a small job of hand expanding leaking tubes in the combustion-chamber of a local cotton mill's aging coal-fired

Danks' Economic boiler, I was homeward bound, my hair, face and hands black as the hobs of hell.

Trundling up steep Hollin Lane, the Austin Devon not at all happy, with much back-firing, the hill climb becoming slower and slower, thus I decided to pull into the Red Lion yard. I also hoped of meeting up with Carrot Crampthorn and the other Boiler Makers. However, I couldn't see any of Phoenix Boiler Maker's lorries or vans, which indicated my mates weren't in the pub.

SUPPIN' WITH CARROT AND THE SQUAD.

I parked the poorly Austin Devon among the hawthorn bushes at the bottom of the pub car park, and blowing some life back into my freezing hands, and although still attired in my coal dust begrimed boiler suit, sporting a dusty tweedy cap on my head, I entered the pub through the back door. The good news to be sure, was that I could see through the glazed door panel that Carrot and Reuben were seated at the bar and supping pints of the 'Swales Swill'.

I could also hear Carrot's deep Lancashire Doric, rattling on about boring football. I quickly opened the tap room door. It was Reuben who saw me first. **"Hell, tis young Alan. Hiya cock, art' thou a millionaire yet?"** inquired the brawny Boiler Maker, grinning. A wee bit of friendly piss taking then went on. **"Can'st lend me five hundred quid Alan?"** Before I could reply, Carrot Crampthorn rose from his bar stool and walked over to where I stood inside the door. **"Hiya Al. Eeh tha looks like you've spent a week inside a ruddy tar barrel."** The Chief Boiler Maker exhorted, his face split into a wide, welcoming grin. **"Dous't want a Barley wine? Now coom o'er 'ere an 'join your owd mates. Teddy an' Paddy are in'th back 'ole playin' darts."**

"My owd Irish pal, Paddy will easily fettle Teddy Tulip" I retorted. Thus, all of us five Boiler Makers thoroughly enjoyed the craic over the next hour. I also looked forward to supping four bottles of the lethal, Barley wine, although this would make me feel a mite unsteady on my feet. Thus, I had decided to leave my tired old Austin Devon van in among the hawthorn bushes at the bottom of the pub car park.

All five of us Boiler Makers shook hands, but just as I was heading for the door, Carrot beckoned me back over to the bar. **"How dou'st fancy weldin' a small patch into the foundation ring area of an owd vertical cross-tube boiler in a Stockport Piggery?"** My eyes lit up, and with a smile I gave my former Chief the thumbs up. Carrot leaned over the bar to help himself to a clean beer mat.

Pulling a pencil out of his shirt pocket, he scribbled down the name and address and phone number of the Stockport pig farmer requiring a Boiler Maker. **"This chap Elijah Wrigley is th' pig farmer's name, and he is a grand feller, a mate o' mi brother-in-law, our Colin. Owd Elijah will pay thee i' pound notes. Good luck lad".**

After the friendly craic and the ale, I left the pub and walked the quarter mile up to my home at Lilac Cottage. After a good scrub in the tin bath in front of the scullery fire to remove the ingrained coal dust, my plan was to walk back down to the Red Lion and drive the Austin Devon the short distance back to Lilac Cottage and to park the sickly old van on the cinder patch close by our front door.

Later, after enjoying my mother's Bury black pudding tea, wrapped up in my ex-Royal Navy duffel coat, kindly gifted to me by my old Master at Phoenix Boiler Makers, owd Jemmy Shuttleworth, in the biting cold, I trudged down Hollin Lane to the Red Lion car park which was in pitch blackness. Using a small battery-operated torch, I unlocked the Austin Devon's cab. I then spent the next freezing half hour attempting to coax life into the van's stricken engine. However, the more I cranked the starting handle, the engine remained lifeless.

Hell's Teeth I thought. Without the van I cannot travel down to Mr. Wrigley's pig farm near Stockport. I felt extremely tired, and now with the effects of the Barley wine wearing off, and my concerns regarding the broken-down van, I felt rather low. The trudge back up the hill on the icy flagstones, the cold seeping into my back and my jean-clad legs, made me feel, even more miserable.

However, once back into the warmth of Lilac Cottage, where mother had brewed a large teapot of Hornimans, my mood lifted. Whilst supping the brew I got stuck into a plate of freshly baked Eccles cakes. Mother bade me goodnight and went off to bed.

The next morning dawned much colder, for it had snowed during the night to blanket the village flagstone footpath with three inches of the white stuff. Notwithstanding, the freezing conditions, I trudged the couple of miles down Hollin Lane, then along Rochdale Road and up to St. Leonards Church in Middleton, where after climbing the steep hill and moving through the snowy grave yard and then descending on the other side, I soon reached Boarshaw Road and my little Boilershop.

Once inside the tiny office I had built from old timber packing cases, I picked up the telephone receiver and dialled Mr. Elijah Wrigley's phone number. A friendly male voice answered, **"This is Stockport 9559, Edgeley Marsh Piggeries. Mr. Elijah Wrigley speaking."**

After mentioning Carrot's brother-in-law Colin and introducing myself, and with much embarrassment regarding my transport problems, I stated my willingness to repair his boiler. **"Nay, young fella, there isn't a problem for after I've fed my 600 pigs, I shall motor up to Middleton in my van. Me and thee will load up all of your tools and tackle and I'll drive you back down here to our Piggeries on Edgeley Marsh.**

Now just give me some directions. It's no bother at all." The kind voice assured me.

Whilst I patiently awaited the arrival of Mr. Wrigley, I carried out a number of minor welding and fabrication jobs for the Middleton & Tonge Co-op, who was one of my best paying regular customers. I also checked three of my tool boxes to ensure all was in order for the job ahead.

However, when Mr. Wrigley eventually arrived in his dark blue Morris LD van, it was after three o'clock and the winter night had started to draw in. After exchanging pleasantries, I opened the van's side door when I noticed there were perhaps a couple of dozen, much battered galvanised steel bins that reeked terribly of fresh pig muck and swill. I wrinkled up my nose at the offensive stench.

Mr. Wrigley was indeed a most pleasant chap aged around the lean side of 60. He had a ruddy complexion and cheerful disposition. Together we quickly loaded up my heavy tool boxes. Mr. Wrigley had offered me the loan of his own electric-arc welding set and oxy-acetylene burning gear which I had thankfully accepted.

Thus, we set off down through Middleton and along Manchester New Road until, after six miles we came to Manchester Piccadilly Railway Station; from here we took the Stockport Road and after driving for around six or seven miles further, Mr. Wrigley then drove the van down a narrow snow-covered lane and into a large cobbled yard with rows of clean and tidy-looking pig-sties. Two powerful electric lights illuminated the scene.

THE LUMBY VERTICAL CROSS-TUBE BOILER REPAIR.

I quickly dismounted from the cab. Much relieved to escape the powerful stench of rotting pig muck that had pervaded the small space from the van's cargo area in the rear which contained the bins. Despite being clad in my thick boiler suit beneath a heavy chrome-leather welder's jacket, my scorched and holed tweedy cap perched upon my head, nevertheless, I felt extremely cold. **"Right ho, young fella me lad, this is the Boiler House",** said Mr. Wrigley as he pushed open the green painted door. The room inside, illuminated with two bright electric light bulbs, appeared neat and tidy, just like the buildings out in the yard. I walked over to the 13 feet high, vertical Lumby cross-tube boiler that was in the centre of the room, and shone my torch around the foundation ring. This revealed a small, localised area of obvious historic leakage around three of four rivet heads.

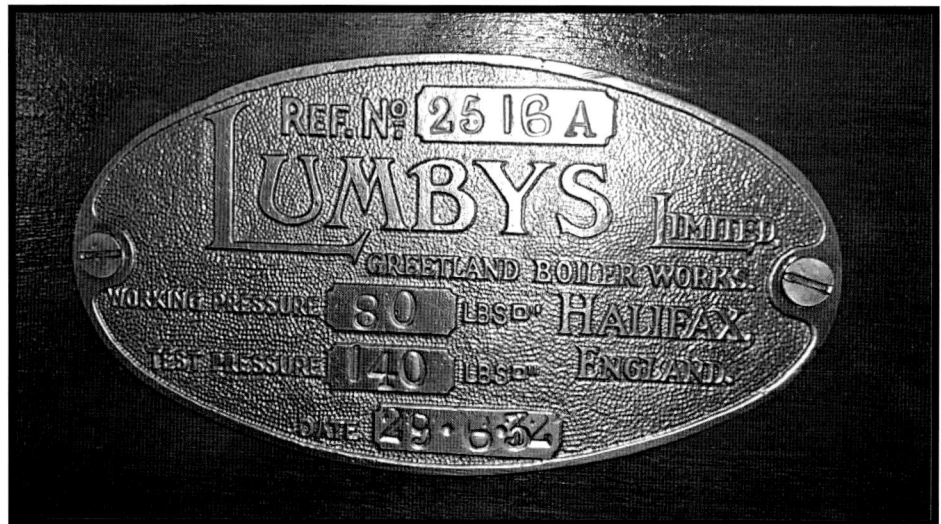

Mr. Wrigley had previously explained where the boiler was leaking from, which was a small section of the plate forming the ash-pan, beneath the line of rivets being wasted. Looking and musing about the work, I decided I would make, fit and weld-in a small patch into this non-pressurised area of the boiler. The slightly leaking foundation ring rivets I would seal by hand caulking. I noticed also that the blowdown valve was in the open position therefore, the boiler had been blown-down and was empty.

"Well, young fella, can you make my old boiler steam tight?" asked Mr. Wrigley. After assuring the genial, red-faced Pigger that I would indeed soon fettle his boiler, he then showed me the welding set and cables which were covered by a tarpaulin sheet in the corner of the Boiler House. I also noticed, Mr. Wrigley's oxy-acetylene burning gear close by. I would require this equipment for burning out the small section of wasted plate. ***"I'll go and make us a brew."*** Said Mr. Wrigley.

I quickly unloaded my ex-British Army ammunition boxes containing my hand hammers, cold chisels and special caulking chisels, plus a packet of Rockweld Vertend 10 s.w.g. welding electrodes, my head shield and leather gauntlets from the Morris LD van. I brought it into the Boiler House and laid the tools and other items close to the Lumby boiler.

The interior of the Boiler House, felt considerably colder than out in the yard, so to keep warm, I got stuck in setting up my own tools and then connecting Mr. Wrigley's welding set earth lead onto the boiler, and then throwing the switch on the 500 volt switch box mounted on the wall.

Next, I replaced the copper nozzle on Wrigley's B.O.C. burning torch with a fresh one straight from my tool box. I was now ready to commence operations.

I hammered off the loose rust scale from around the affected area and then wire brushed the metal wastage. I then drew French chalk lines, followed by centre-popping. Donning my ex-Phoenix 'frog goggles', I sparked up the gas-axe. I applied the torch and cut out the wasted material resulting in a neatly burnt hole. After chiselling off the slag, I cut weld preps onto the plate edges with hammer and cold chisel. The next operation was the cutting of the small 4 inch by 3 inch by ⅜ inch patch from the piece of boiler-quality plate I had brought from my Boarshaw Boiler Works.

I had just completed burning out the patch plate, when my new friend, Mr. Wrigley appeared carrying a large steaming mug of tea and a plate of arrowroot biscuits. ***"Here you are lad, the tea will warm you."*** It was now 6.30 in the evening. The burning and chiselling work had actually warmed me up, and the tea and biscuits were most welcome. Earlier I had noticed, that the boiler man-lid jointing ring had been 'slavering', thus I elected to replace it with a new one. One of my tool boxes contained a large quantity of asbestos joint rings for man-lids and mud-lids.

Propped against the Boiler House wall was a stout wooden ladder, which I would use for climbing up to the man-lid. But first things first, I thought. Firstly, I will fettle the welding in of the plate patch and caulk the rivets. Thus, I soon had the small patch neatly welded into place. Then upon selecting a small caulking chisel, I proceeded to caulk several of the foundation ring rivet heads close by the patched area. Mr. Wrigley then re-appeared and upon donning a pair of thick-lensed spectacles, and walking over to the boiler, knelt down low and carried out a close visual inspection of my work. Looking up towards me, a huge grin lit up his ruddy face. **"Eeh young fella, you've done me right proud. My old boiler should now last me out**." Exhorted the genial Pigger. I felt very pleased with myself.

My next task was to change the leaking man-lid joint. **"Can I leave you to finish, I must go home to check on my wife who is in a wheelchair. Our house is about a mile down the lane but I'll be back in about an hour."**

I smiled an acknowledgment and gave him the thumbs up. He then handed me a parcel of fish and chips wrapped in several layers of newspapers. **"I have just been down to the village chippie for your tea."** Then, with a wave, he left me in the frigid, silent Boiler House. Despite the earlier intake of arrowroot biscuits, I was, nevertheless, extremely famished, and soon wolfed down the delicious tasting fish and chips.

BADLY BRUISED WRISTS.

Standing on the wooden rungs of the ladder propped up against the Lumby boiler, I carefully removed both 1½ inch Whitworth nuts securing the man-lid. Carefully, I then eased off each of the heavy cast-iron dogs. I placed both dogs on the domed boiler top above my head. Delivering a sharp belt with my hand hammer to the centre of the man-lid, while holding on to the left-hand threaded spigot, thus I managed to break the seal.

I pushed the heavy man-lid inwards and rested it onto the firebox crown. I then removed the old asbestos joint, cleaned and wire brushed the 'land', applied manganesite, and fitted a new asbestos joint. My next job, involved using the conveniently sited hose pipe to fill up the boiler with clean water. I also added two shovels full of water treatment chemicals that Mr. Wrigley had earlier instructed me to use.

After turning on the low-pressure water supply, I packed up most of my tools, unplugged the electric-arc welding set, wrapped up the burning gear and generally made ready for my departure. I could hear the slow trickle of water filling the boiler. Many boiler repair jobs, unfortunately involved lots of waiting time. So, after rummaging around, I picked up a couple of Pig Farming magazines, and wrapped in my chrome leather welder's jacket, I settled down to await the filling of the boiler. It was over an hour later when the water had reached the half way mark on the glass tubes of the water gauges. **"Hells Teeth! I muttered. Where is old Wrigley?"** I felt extremely cold, tired and low spirited.

Nevertheless, it was time to tackle the refitting of the man-lid into the boiler. Thus, up the ladder I went. Reaching in with both hands through the man-hole I grasped the threaded spigots and lifted and coaxed the heavy man-lid into position in the shell.

Unfortunately, however, the combination of extreme tiredness and feeling chilled to the bone, caused me to slightly relax the grip of the man-lid, which suddenly fell backwards onto the fire box crown, at the same time trapping both of my wrists onto the sharp edge of the compensation ring. Despite pains shooting up both arms, I daren't let go of the man-lid, for it could have fallen to become jammed down in the water space between the shell and the fire box. Thus, I was trapped, and by now feeling rather unwell.

The pain in my wrists was getting worse. And where the hell was the old Pigger? I pondered. It felt that he had been gone for hours. To ease the pressure on my rapidly swelling wrists I gingerly moved up to climb a higher rung of the ladder. I started to shout loudly for help. I had now been in this painful situation for about a half hour. I shouted again, my throat becoming hoarse. Suddenly, I heard the Boiler House door open.

"Hello, hello," I then noticed a torch beam lighting up the gloom of the Boiler House. Twisting my head around, I then made out the owner of the voice. It was a police constable. I urged him to climb up the ladder behind me and to assist with the lifting of the man-lid to relieve my almost dead, painful wrists. Fortunately, not only did the officer valiantly free my wrists, but he also assisted with the fitting of the man-lid and the tightening of the nuts.

A few minutes later a rather troubled Mr. Wrigley entered the Boiler House. Him and the constable were well acquainted. He then went off and quickly returned with mugs of tea and a large plate of arrowroot biscuits. I gave the kindly constable a large dollop of Swarfega which he used to cleanse his filthy black hands.

We then shook hands, I thanked him and he bade me a hearty goodnight. Another hour later and with Mr. Wrigley's help I managed to steam the boiler to my ultimate satisfaction. It was well past midnight when Mr. Wrigley pulled up the Morris LD van outside Boarshaw Boiler Works. We quickly unloaded the tool boxes and placed them inside the Works. The genial Pigger then dropped me off very close to my home at Lilac Cottage at Top of Hebers.

The next morning, I posted Mr. Wrigley's bill which amounted to £24. 7s 6d. Mr. Wrigley kindly sent me a cheque with a very complimentary thank you note by return of post.

Around dinner time, and feeling dog tired, I called in to see Gordon, the local milkman. I told him about the previous evening's adventure involving me being rescued by a genial Stockport police constable. I then asked Gordon who was locally well known for dabbling in second-hand vans and trucks, if he could find me a suitable, strong and reliable work horse of a van for hauling around my heavy boiler making equipment and tools. Quick as a flash he motioned me to follow him into one of the large old barns. He pointed to a most pristine-looking Morris J4 van, and explained that he had only just acquired the van from a Manchester cardboard box manufacturing firm, who clearly had taken significant care of the vehicle, keeping it in decent fettle.

Gordon then invited me to take the Morris J4 out for a run up to Heywood and back. The van drove beautifully, and on my return, following a little bit of price quibbling on my part, I purchased the van for the then quite considerable sum of £90 cash. Gordon took my old knackered Austin Devon away for scrap.

Within a week of the Morris J4 becoming my property, I instructed a local sign writer to paint in large yellow letters on the van sides and rear door:

H.A. McEWEN (Boiler Repairs).
Boiler Makers and Welding Engineers
Boarshaw Boiler Works, Middleton, Lancashire.
Telephone MID 6008

My badly bruised wrists were extremely painful for around a fortnight.

An advertisement for Spencer-Hopwood Vertical Boilers.

A Vertical Cross-Tube Boiler of unknown make.

Chapter 3

MY RETURN TO DANDELION COLLIERY PIGGERIES.

The irascible, yet basically decent old Pigger, Hiram Jackson of Dandelion Colliery Piggeries, subsequent to me renewing his vertical cross-tube boiler's uptake tube was evidently so satisfied with my boiler-making skills, that he had promised to spread my name far and wide. And indeed, he did, for Hiram had kindly given my name and phone number to a fellow Pigger in need of urgent boiler repairs who also operated on the Dandelion Colliery Piggeries site.

Thus, one sunny, yet cold and windy morning in January 1970, after receiving a phone call from one Harry Owler, who kept around 300 pigs, resulted in me paying a second visit to Silly Country and the former Dandelion Colliery site.

As I drove my red-painted Morris J4 van down the familiar cinder track leading into Dandelion Colliery Piggeries, my imagination conjured up the frightening image of me falling off the Jackson Brother's Boiler House roof and landing in deep, freshly deposited, reeking pig muck, where I was savagely attacked by a nasty tempered monster-sized sow. Indeed, this non-too pleasant occurrence had happened just about three months earlier.

Back then, I had hoped and prayed that I would never have to visit stricken Dandelion Colliery Piggeries ever again. However, here I was driving my Morris J4 van into the wilderness of countless pigsties, the air filled with the relentless sounds of squealing pigs; and the appearance of the wild-looking, scruffily attired characters known as the Piggers.

As the Morris van chugged along the rough cindery track in second cog, while using my hand to shield my eyes from the brilliant piercing rays of the low, winter sunshine, my gaze travelled downhill into the shallow valley bottom shared by the railway line and the disused rubbish-choked canal; for I could clearly see glinting in the brilliant sunshine an astonishing hotch-potch of ancient-looking, crumbling brick pigsties, numerous ram-shackle timber huts and sheds and scores of stubby, red-rusty, iron smokestacks thrusting skywards through rust-eaten wriggly tin roofs of what were obviously Boiler Houses; used for boiling pig swill.

"Hells Teeth!" I muttered, while sucking onto a strawberry-flavoured Spangle sweet. ***"Beneath yon owd corrugated roofs, there could be scores of vertical cross-tube and other boiler types requiring my boiler-making skills"***. Fascinated with the bizarre panorama before my eyes, I killed the Morris's engine. Here and there I could discern wisps of white steam rising into the cold wintry air. From the tops of several iron chimneys plumes of black smoke issued forth.

As I drunk in the sights and sounds of the ubiquitous workings of pig-rearing, pig-feeding, and other Piggery duties going on inside the extensive groupings of Piggeries, my nose started a-twitching; for richly pervading the January air was the rather unpleasant aromas of boiling pig swill. What a rather bizarre place to visit: a throwback to much earlier days, I thought.

Firing up the Morris, I continued the crawl down the undulating cinder track, in the attempt to locate the holding belonging to my new customer, Harry Owler. While passing what were clearly recently-built pigsties on my right, I spied a balding, elderly chap wearing moleskin trousers, accompanied by a very buxom, pretty lady of middle years. They were attempting to hoist a long length of wooden telegraph pole onto the top of a building under construction.

BENEVOLENT BILL, THE YORKSHIRE ENGINEER.

Upon seeing the bald chap and the attractive lady clearly struggling, I quickly stopped the Morris, and after exiting the vehicle, I jumped over the low timber post and rail fence and without even catching breath grabbed one end of the telegraph pole. Together, all three of us pushed the pole upwards and into the desired position on the roof.

"Well, thanks a bundle young man" said the bald, elderly gentleman, whom I detected spoke with the nasal, yet friendly accent of the West Riding of Yorkshire, while mopping his glistening sweaty brow and pate with a lump of rag. **"This is my wife, Elizabeth, or Bessie".** He said, wiping his hands on the sweat-soaked rag. **"My name is Bill. How do you do?".** All three of us shook hands, which felt very formal considering our surroundings. I introduced myself and then asked Bill and Bessie if they could direct me to Harry Owler's Piggery. They quickly pointed out a tumble-down clutch of buildings situated about a further 100 yards down the cinder track on the left-hand side.
"Ah young Alan, I can see from the painted sign on your van that you a Boiler Maker. Are you seeking work?" Bill

politely inquired. I then proceeded to explain to the friendly Yorkshireman and his lovely wife, that even though I was keeping the wolf from the door by carrying out welding repairs to lorries and tar-melters; and not least to woebegone steam boilers owned by pig farmers, nevertheless, I yearned for a return to regular boiler-making work on industrial sites.

"Well Alan, when I wear my normal hat, I am the Group Engineer of a large, nationally renowned engineering company. Under my care we have four, 16,000 lbs. per hour Ruston & Hornsby, 3-pass, wet-back steam boilers. One of the boilers requires a full re-tube and welding work in the furnace tube. Would you be interested in taking a look?"

Bill then rummaged in his shirt pocket and fished out a small, white business card which he presented to me. Displayed in neat copperplate black ink was his company's name and address which was located near Stockport. Underneath, Bill's full name was printed together with the letters: C.ENG.M.I.MECH.E. which I knew meant that Bill was a chartered mechanical engineer, and a member of the famous Institution of Mechanical Engineers.

"Yes Bill, thank you." I gratefully replied. Bill and Bessie then informed me that pig-breeding was a recently acquired past-time, which was enjoyed by both, as it was a total change, especially for Bill. Bill planned to retire from engineering in two years-time thus, the couple intended to make pig-breeding their full-time business.

The following Tuesday, as arranged, I met Bill at his company engineering firm, and after viewing the proposed boiler repair work, and tendering a budget price, the official order was granted to me. However, this exciting story will be told in a future book.

Bidding Bill and Bessie a friendly wave from the cab of the Morris J4, I then drove slowly down the cinder track until, I arrived outside the ramshackle holding belonging to Mr. Harry Owler. I quickly alighted from the cab, and was immediately greeted by a tall, muscular chap aged about 45, with a shock of greasy blonde hair.

He was clad in a much-patched pair of pig muck stained bib and brace overalls and a red scotch-patterned shirt nestling beneath a ripped and filthy-looking dark-blue donkey jacket. On the man's feet were a huge pair of stout army boots, tied with string in place of laces. *"Hello there, young man. You're bloody late! I saw you stop and help yon posh Yorkie an' his woman. You should have been here mendin' my old boiler much earlier. I now want you to get crackin' on or, I'll be knockin' a lump o'brass off your bill. And by the way, I'm Mr. Owler."*

Hells Teeth! – what a greeting, I thought. *"Right young fella me lad, follow me."* Mr. Owler, kicked open a woodworm ravaged double door set into the wall of a brick building, and with a wave of his hand motioned me to follow.

A COLONIAL HORIZONTAL STEAM BOILER.

The building's interior measured approximately 20 feet long by 15 feet wide. A row of small windows set high on the rear wall admitted sufficient light for me to see that stacked around the walls, were huge piles of freshly sawn fire-wood billets, cut from rough old demolition site timber. Set into the middle of the Boiler House was a rare beast of a Colonial, horizontal steam boiler.

Brass Maker's plate from a George Sweetlove Ltd Steam Boiler.

Externally fired Colonial Boiler.

A red rusty, riveted steel chimney mounted atop the smokebox rose up through a jagged hole in the wriggly tin roof. *"Well, this is me old boiler. She's a bloody good steamer, but is reet thirsty on wood fuel. By the way, what do they call you? What's your name?"* I introduced myself, offering my hand, which Harry Owler eagerly grasped in his huge callused right hand. *"My name's Harry. Harry Owler. You can call me Harry. Thanks for traipsing down into Dandelion. I know it is a fair distance from Middleton. I am really sorry for rantin' and moanin' earlier. It ain't an excuse, but with the boiler being down, I've not cooked any swill for over 24 hours, and my 300 hungry porkers have already eaten a week's supply of very expensive corn grub, which I cannot afford."* Explained a most chastened Harry as he sat down on an upturned swill bin.

While Harry was unloading his troubles, I was enjoying myself giving the George Sweetlove of Bolton-built Colonial boiler a good dose of looking over. Upon noticing a tarnished brass plate bolted onto the smoke-box door, while still listening to Harry, I gave it a rub with my thumb. The boiler was built in 1936 for a maximum steam working pressure of 80 p.s.i. I asked Harry how long he had owned the boiler. He explained that he had taken over the Piggery from his recently deceased father, who had acquired the boiler second-hand in the late 1940s.

"The old boiler has always performed well, for as long as I can remember. About two days ago, I happened to notice white steam billowing out from the chimney top."

Harry then explained the following: *"After letting th'fire burn out, and giving the boiler time to cool down a little, I operated the blow-down valve to drain down the water as it was flowing out of the front of the furnace. I unfastened the catches on the fire door casting in the hope of finding the leak. I then saw that the end of the furnace had cracked. In fact, it was a bloody great crack! No boiler meant no swill boiling, so I went across to see that miserable old sod Hiram Jackson. It was Hiram that gave me your name and phone number."*

While Harry was engaged chatting, I picked up a large adjustable spanner and got stuck in to removing the half dozen ¾ inch Whitworth nuts that held in place the heavy fire door frame casting. Harry assisted me to carry in my heavy tool boxes, the Pickhill oil-cooled electric-arc welding set, welding cables and also a set of pull-lifts and sling-chains. Looking up, I saw a stout timber roof truss conveniently located just above the boiler's front end. I then asked Harry if he had a ladder. I then commenced to set up the sling-chains in readiness for the lifting of the fire door. Harry re-entered the Boiler House carrying an 18 foot ladder.

He placed the ladder against the roof truss and picked up the end loop of one of my long, steel chains. Up he went to wrap the chain around the truss. I passed him another, shorter chain, which he quickly connected onto the first chain. Thus, the chain now hung about two feet above the fire door frame casting. I attached the pull-lift, and by operating the lever took up the weight of the heavy casting. Using a small pinch bar, I then eased the fire door frame casting away from the boiler's furnace mouth and lowered it onto the floor.

A vigorous wire brushing, followed by sweeping with my hand brush removed the consolidated mixture of ash, coal dust and lime scale from the base of the riveted furnace tube to front end plate attachment. I then noticed a circumferential crack of about 1/16 inch running from the base of the furnace centre line up to around 3 inches on either side. Because previously, I had been involved in numerous repairs of this type of cracking – resulting from internal 'grooving' – I understood therefore, that a temporary repair could be readily executed.

TACKLING A BIT OF GROOVING.

Yet before I commenced tackling the grooving defect, I went to the rear of the boiler and fully opened the blow down valve for several minutes, which dropped the water level well below the defect.

The next stage involved me plugging in my Pickhill electric-arc welding set into a conveniently sited 13 amp, 240 volt socket fastened to the wall. Opening up the lid on my 'chisels' tool box I selected three bull-nosed chisels. From within the 'hammer' tool box I took hold of a 3 lb hand-hammer.

After chasing the line of the 'grooving' with French chalk, using the narrow-tipped bull-nose chisel, I then commenced chiselling out the crack, removing the metal right through into the water space. This process was followed by cutting wide weld preps using the other bull-nosed chisels. A close visual examination, proved as I had originally thought, that the water side of the furnace tube was extensively 'wasted'. There were also significant hard scale deposits adhering to the water side of the boiler furnace and end-plate. Nevertheless, my immediate task was to expertly repair the grooving resulting in the boiler eventually returning to boiling Harry's pig swill.

With the chiselling completed, I donned my head shield and with a Rockweld Vertend 12 s.w.g. welding electrode placed in the electrode holder, I started to carefully lay down a neat root weld in the bottom of the weld prep. This was a slow process due to the severely corroded plate and the heavy scale on the water side, but I persevered, and after completing the root run, using the middle-sized bull-nose, I chiselled off the high spots on the weld. This was followed by 'peining' the weld by hammering the surface with a special shaped peining chisel.

All was going fine, that is, until I blew the 13 amp fuse on the wall-mounted switch box. Locating Harry, who was out in the yard scrubbing swill bins, I got him to show me an alternative electric socket which appeared to be quite new. It was however, unbelievably, located inside one of the larger pigsties. Harry cautioned me, advising me to watch my back when entering the sty, for the pigs, having gone without a feed of boiled swill for several hours, were by now very hungry and extremely agitated.

I assured him that I could well look after myself and wasn't scared of pigs. Thus, I went out to the Morris van and returned to the pigsty with a long electric extension cable. To insert the electric plug into the wall socket, I had to clamber over a low brick wall and into the pigsty.

I had just managed to push the electric plug into the wall mounted socket, when the air was rent with a cacophony of load grunting noises. Suddenly, I was surrounded with about a dozen fat, angry porkers, who appeared to take a fancy to me. One of the sows, perhaps braver than the others, attempted to take a bite out of my boiler suit-clad legs. Hells Teeth! What have I got myself into I thought.

Although I wasn't in the least frightened, for I was a country lad, who felt safe with farm animals, and particularly pigs. Then however, another much larger pig managed to painfully nip my thigh. It hurt like hell, and I knew I had to get out of the sty. Just as I was about to mount the wall, several pigs launched themselves forward, their wide-open jaws revealing rows of huge, yellow teeth. Fortunately, tucked into my leather belt, was my trusty 3 lbs hand hammer. Quickly I yanked it free and brought it down hard on the head of the nearest wild pig. The animal squealed blue murder and raced away followed by the others. Thus, I escaped being eaten alive by a herd of Dandelion Colliery Porkers.

By late afternoon, the welding was completed, and after removing the safety valve, I stuck a hose pipe into the opening to fill the boiler, thereby allowing me to hydraulically test the welding repair. Harry brought me a brew of tea and an old folded newspaper containing Jaffa Cakes. **"Just like old Hiram Jackson, I also remove waste biscuits and butter fat from the local biscuit works."** Said Harry, happily munching on a gob full of biscuits. After the brew and biscuits, I tightened the bolts holding down the blank flange on the safety valve orifice, and using the electric boiler feed pump I conducted the hydraulic test which was proven satisfactory.

Harry kindly chose to assist me once more, for while I was refitting the safety valve, by using the pull-lift, he managed to refit the heavy fire door frame casting and tighten up the Whitworth nuts. I then opened up the blow-down valve, to drop the feedwater to the normal working level. Harry and I fitted the fire bars into the furnace, which was followed by the friendly Pigger placing an armful of thinly chopped sticks onto the bars. He then inserted some diesel-soaked cotton waste and applied a lit Swan Vestas match. The wood flared up. Harry kept adding small lumps of firewood until, the furnace was full of blazing fuel.

He then closed the cast-iron door and opened up the air-intake regulator.

Meanwhile, I packed up my tool boxes and welding cables, then rolled out the Pickhill welding set close to the Morris van. Harry kindly assisted with the lifting of the heavy welding set into the back of the vehicle.

I suddenly began to feel really tired. I had been working at Harry Owler's Piggery for around 7 hours. Sitting in the cab, I wrote out my bill which amounted to £13. 10s. and handed it to Harry. With a smile, he disappeared into the Piggery to return a couple of minutes later to hand me £15 in cash. **"Have a couple o' pints of ale on me Alan. You have done me proud."** I handed him a few recently printed business cards. We shook hands.

Looking back, I really did enjoy repairing both Jackson Brothers and Harry Owler's boilers and of course meeting Bill, the affable Yorkshire Engineer at Dandelion Colliery Piggeries.

A Pigger happily steaming a swill bin.

Chapter 4

WORLD'S END PIGGERIES JOB.

Looking back, I now fondly recall those colourful, eccentric, rumbustious characters who added much excitement into my daily toil: indeed, the Piggers belonged to a different age.

BART and NUALA O'GRADY, CHAMPION PORK FATTENERS, and MAGGOT BREEDERS.

One such character was Irishman, Bart O'Grady who evidently had been born and reared in a turf shack in the wilds of Connemara, Co. Galway. Bart was in his late fifties and lived with his long-suffering, pretty wife Nuala in what could only be described as a semi-derelict, smoke-blackened, double-fronted stone Georgian farm house, from where the couple operated their main business, World's End Piggeries, which produced fat, high quality pigs. The Irish couple also specialised in breeding maggots for anglers.

The O'Grady's were childless, but had three huge, hungry-looking Alsatians as pets which fortunately were kept locked up during the day. The O'Grady's old farm house was set on an elevated, west facing slope of sparsely-wooded Blue Lees Clough, overlooking Oldham Edge.

Bart's favourite tipple was Paddy Irish Whiskey. He was well known for draining a full bottle during the regular poker sessions he and his fellow Piggers enjoyed in the farm house parlour. Local gossip was, that upon Bart receiving the Co-op cheque for the lorry load of well fattened, high quality porkers he had sent down to the pork factory in Cheshire, he would together with his cronies, delight in taking the train to either York, Thirsk or Redcar races, where, while supping from a bottle of Paddy, his bulging wallet would be rapidly drained due to him placing bets on lame duck horses. Late the same evening, he would arrive back at World's End, strongly reeking of Paddy whiskey, his wallet now empty. The hapless Bart would then receive a verbal hammering from Nuala.

Bart was powerfully built with bulging arm muscles, no doubt gained in his youth, after arriving in the North from Ireland and working as a bricklayer's hod-carrier on construction sites all over Lancashire and Yorkshire. Despite him working in what could only be described as pig muck and reeking pig swill, Bart was a rather natty dresser. On his head he wore a smart trilby, complete with feather. His muscular legs were protected by thick, green Donegal tweed trousers, and his jacket was also of the finest Donegal tweed. His feet were shod with a highly polished pair of heavy, iron-shod brown leather boots. Clamped between his tobacco stained teeth there was usually a glowing Harlequin cigar. Bart was as tough as Irish bog oak; he was usually a rather happy-go-lucky, warm hearted, kind character. However, he had a rather sensitive disposition and if crossed then he could quickly rise to become extremely angry.

Nuala was just a tad younger than her husband, but her beautiful shoulder length, curly auburn hair, warm and passionate piercing green-blue eyes, and wrinkle-free fresh Connaught complexion gave her the appearance of a woman fifteen years younger. I later learnt from the maggot lad, Hughie – whose job it was to encourage the weekly production of prodigious quantities of maggots much in demand by anglers across the British Isles, that his boss Nuala, was the daughter of Irish gypsies. She had been born in a bow-topped, horse-drawn caravan in a magical, remote valley campsite in the shadow of Nephin, the 2646 feet conical-shaped mountain in Co. Mayo.

Nuala was highly intelligent; a fluent speaker in both Irish Gaelic and English. She was also proficient in writing and was an extremely adept business woman. Without her highly successful, profitable maggot business, then money would have been tight. On the many occasions that Bart would return home drunk and penniless, and with the temper of a raging bull, he would have frightened most women. However, Nuala would soon calm him down and due to her intelligence and quick wit, she could verbally hammer Bart when he was being vocally aggressive. It was clear however, that Bart deeply loved Nuala and also respected her highly. Even when full of whiskey and prone to much shouting and swearing, Nuala would utter a few gently words in Irish which resulted in her husband becoming quiet and peaceful. She would then serve him strong coffee and perhaps a couple of servings of thick, salty oat porridge.

One pleasantly sunny morning, I was busily engaged gas welding a small steel patch I had formed on my anvil to fill an unsightly hole in the front nearside wing of a Bedford 8-ton lorry belonging to my neighbour, Alec, the local coal merchant, whose yard was sited next to my Boarshaw Boiler Works. The lorry was parked up in front of the Boiler Shop. Upon me hearing the shrill ringing tones of my newly installed telephone, I quickly turned off the British Oxygen Company welding torch and picking up the telephone receiver within my tiny office, was greeted by a friendly Irish voice explaining that his 'auld' ship's steam boiler had developed a serious water leak resulting in the boiling of much needed pig swill coming to a standstill. The jovial Irishman then informed me that his wife had in fact jotted down my business name and phone number upon her recently seeing my Morris J4 van parked outside an Oldham cotton mill.

After assuring the Irish pig farmer that I could inspect his boiler straight after dinner time and upon writing down details of his address and location, I then returned to my gas welding duties on the coal lorry. Upon completion, I used an emery block to smooth down the weld metal of the newly welded patch. I then applied a thick coat of black paint. Firing up the large, Bedford petrol engine, I then delivered the lorry into the coal yard.

Chomping on a cold bacon muffin, and swigging from a flask filled with Hornimans tea, I drove the Morris J4 van single handed, as I trundled up Yorkshire Street in Oldham, en route for Mr. O'Grady's pig farm at Blue Lees Clough. Following the directions I had jotted down onto a writing pad, and after slowly ascending the steep moorland road for a distance of around eight miles heading into the Pennines, I suddenly came to a painted sign with a large arrow: O'GRADY'S WORLD'S END PIGGERIES. I stopped the van to drink in the surrounding breathtakingly, beautiful Pennine scenery. The large, sprawling cotton town of Oldham, together with its numerous, mammoth red brick mills and skywards thrusting chimneys was spread out below like an ink splodge on blotting paper. Further still, the stronger early afternoon light revealed clearly to the south west the gigantic reach of Manchester and Salford.

Moving off, I followed the direction of the painted arrow up a stony track to finally emerge into an extensive farmyard built on a stony plateau. Forming a quadrangle were ancient looking barns and rows of pig sties, built from the solid, local Pennine sandstone. At the far end of the yard stood a large, double fronted stone farm house which surprisingly, appeared to be totally surrounded with great numbers of battered steel bins. Parking the Morris, I alighted and walked across the cobbled yard towards the farm house. Approaching the steel bins, I noticed they had plywood lids. Inquisitively, I lifted the lid of the nearest bin, and then wished I hadn't been so nosey, for the air was immediately strongly pervaded by the highly offensive odour of putrefying flesh. Down in the bottom of the bin was a huge swarm of revolting-looking, gigantic blue bottles laying their eggs in a disgusting mess of greenish, stinking, rapidly rotting lumps of cow flesh. Hell's Teeth! – I thought, as I battled against the urge to vomit.

I then realised that the sea of old dust bins was probably brim full of the terribly reeking rotting meat and crawling with huge flies and maggots.

"Good afternoon, young sir. You will be the boiler man?" Lifting my gaze from the dust bin horrors, standing around 20 feet away was a very attractive lady. With the reek of putrefying flesh in my nostrils, while coughing, I waved my hand and uttered a friendly greeting.

"Er …… yes, I am ……. cough, …… cough. Yes, I am Alan McEwen, the Boiler Maker from Boarshaw Boiler Works."

"Hello Mr. McEwen. I am Mr. O'Grady's wife. You can call me Nuala." Replied the softly spoken, pretty Irish lady, flashing brilliant white teeth and a huge welcoming smile. She offered her hand which I had the pleasure of taking.

"Good afternoon. The surrounding views of the hills and moorland, and even dirty old Oldham, from up here at Blue Lees Clough is amazing." I said.

"I have just prepared plates of ham and mustard sandwiches, and brewed tea. Will ye join Bart and myself in the kitchen?"

"Er ….. I've actually had my lunch, ….. er ….. but a sup of tea would be very welcome." Was my reply. I certainly couldn't stomach food at this moment, the sight of the bin contents filling my thoughts. As I followed Nuala through the sea of steel bins, upon her seeing a gangly, ginger-haired youth carrying a fork coming from within one of the pig sties, she shouted a friendly instruction to the lad.

"Hi Hughie, wash your hands an' join Bart and Alan, the Boiler Man here, an' myself for grub and tea." The revolting steel bins were everywhere, right up to the kitchen door. Nuala didn't seem to mind them, or the swarms of blue bottles and the heavy offensive reek that appeared to hang in the air. I followed her into a large flagstone floored kitchen that, due to the small mullioned windows and the low beamed ceiling appeared quite dark. Nevertheless, all appeared to be clean and tidy, with a large old oak dresser with four shelves upon which was displayed a large collection of blue china plates and mugs. Suddenly, a dark oak door opened, and a tall strong looking, tweed coated man walked in.

"Ah, love of my life. So, you've woken up. Bart, this is Alan the Boiler Man. Alan may I introduce my dear husband, Bart O'Grady, Lancashire's finest swine herd."

I walked over and shook hands with Mr. O'Grady. Nuala then bade us to sit around the huge, heavy planked dining table. She then poured out the tea into the blue mugs off the dresser. She placed a huge pile of neatly cut ham sandwiches stacked on a wooden bread board onto the table. The kitchen door opened and in walked the ginger nutted lad Hughie, who then came and sat at the table.

"Now, to be sure you fellas will be hungry. I cooked the ham, which was freshly killed, late last night. So ye all can get stuck in. There is also plenty of tea." She then left the room.

I found Bart O'Grady to be immensely friendly. He told me how Nuala and him, had some twelve years earlier purchased the pig farm, and after taking much advice from the old retiring farmer, had overtime steadfastly built up the business. Being intrigued regarding the maggot breeding business, I asked him how he started. He smiled and shrugged and said that maggots didn't appeal to him and the maggot farming business was Nuala's responsibility. He then related the tale about 'the ship's boiler', which I guessed was probably a Scotch Marine boiler, which had been acquired in the early 1950s by the pig farm's former owners from a ship breaking yard in Birkenhead. I must admit, I was so much intrigued and at the same time excited, that I was straining at the leash to see the old Marine boiler which, up here in the Pennines was an extremely rare boiler curiosity.

Thus, with lunch finished, and excitement rising, I followed Bart across the cobble stone yard to head towards a squat stone building with a pitched flagstone roof, out of which rose an old, rusty, riveted iron chimney complete with a China man's conical cap. This was the Boiler House.

Bart's, huge gnarled hand, opened the iron latch on a much-weathered wooden door, and he stepped inside, with me hard at his heels.

The Boiler House looked to be about twenty feet square. The once whitewashed walls now looked rather dingy. The small square windows admitted sparse natural daylight due to the panes being smeared with dead flies and other insects. In the centre of the Boiler House stood a striking-looking old Scotch Marine, double-furnace steam boiler, the overall length was around 9 feet; the diameter was about 7 feet. It looked to be very well-maintained sporting highly polished bronze water gauges and a 12-inch diameter brass, Hopkinson's steam pressure gauge. The lagging I guessed, was of the original blue asbestos which was fortunately safely covered with steel cladding sheets, painted in red lead.

A large 12 ft. 6 in. diameter, triple-furnace Scotch Marine boiler built by Riley Brothers, Perseverance Boiler Works, Stockton-on-Tees for a trawler.

Staring at the smoke box doors, my eyes logged the various details, while my imagination, at most times fertile led me to imagine this ancient, coal-fired Scotch Marine boiler positioned deep down within the stokehold of a steam-powered, little coaster, or tramp steamer, battling the Atlantic Ocean while delivering goods to ports around the British Isles. In my mind's eye I saw swash-buckling sailors, a white bearded, blue coated ship's Captain, the coal-blackened faces of the stokers; and the Engineers with oil cans and spanners.

Suddenly, coming back down to earth, I quickly noticed a brass maker's plate riveted onto the right-hand smoke box door. Using my tweedy cap to rub off some of the tarnished brass revealed: RILEY BROTHERS (Boiler Makers) LIMITED, PERSEVERANCE BOILER WORKS, STOCKTON-ON-TEES. BUILT 1910. MAXIMUM W.P. 200 P.S.I. The works number and some of the information had become illegible.

What a fantastically and profoundly interesting old boiler, I mused. I was delighted and indeed quite honoured to be the chosen one to carry out the repairs to this old steaming workhorse.

Bart then led me to the back end of the boiler, where I could see the combustion-chamber door had previously been removed and laid close by, together with securing nuts and washers. Leaning over to a wall-mounted electric switch, Bart switched on a hand lamp and motioned that I should look inside the combustion-chamber, where I would easily see the source of the leakage. Leaning inside the man-hole, he then clipped the lamp onto a metal bar protruding out of a smoke tube. By now, I was already attired in my boiler suit. I removed my old tweedy cap from one of the pockets and placed it on my head. I then took hold of my inspection hammer – (a small Callon's Toffee-breaking hammer, given to me by my mother's friend Ruth, the owner of an Ardwick sweetshop) and squeezed my head and shoulders through the 16-inch diameter opening which was followed by the rest of my lean body.

Turning around, I knelt up on a plywood board that previously had been placed across the base of the 'wet-back' combustion-chamber. Pulling my small battery torchlight from my boiler suit top pocket, I pointed the beam onto the rows of stay-bolts on the back head. Even though the boiler had been blown down, there was significant evidence of 'scouring' of the plate resulting from serious steam leakage. A considerable number – probably over 100 out of perhaps 800 stay-bolts and nuts had suffered significant metal wastage; several being totally missing. Close inspection revealed the threads on many of the stay-bolts were gone and moreover, the caulking had 'relaxed' – resulting in serious leakage. This was the reason the boiler was shut down. Without a boiler, the swill being delivered daily on a swill supply contractor's lorry, without steam could not be boiled. Thus, Mr. O'Grady's huge herd of pigs were being fed on expensive corn.

Staring at the steam-wasted stay-bolt nuts I tapped several with my small hammer which proved them to be loose. Due to me having considerable previous experience of carrying out similar repairs on marine boilers, I therefore, fully appreciated that the job could, if carried out in the traditional way turn into an extremely large undertaking. For in the old days of boiler-making prior to the introduction of quality electric-arc welding, all of the defective stay-bolts would have necessitated being cut out, a quantity of new, 12 threads per inch stay-bolts manufactured, the stay-holes reamered and re-tapped; the new stay-bolts screwed home, the ends riveted over and then caulked; and new stay-nuts fitted. This would however, turn Mr. O'Grady's boiler breakdown into a huge, very expensive undertaking. Thus, I opted to employ the latest modern welding procedure.

Recollecting a similar job on a Scotch Marine boiler that I had been involved with while working with Carrot Crampthorn's Squad, because I was an experienced and coded welder, the local Boiler Inspector had allowed us to remove the stay-bolt nuts followed by the cutting off of the projecting stay-bolts, leaving just ⅜ inch standing proud. Following the machine peining of the plate surface surrounding each defective stay-bolt to remove scale, I then commenced welding the stay-bolt ends, forming neat fillet-welds onto the surrounding plate. The results were acceptable and the Scotch Marine boiler therefore, was quickly returned to steam. The Boiler Inspector was fully satisfied with the repair. This was the procedure I decided to apply on Bart's Scotch Marine boiler.

Thus, after explaining the electric-arc welding method to Bart and quoting my price of £75, with a smile and a thumbs up, Bart gave me the go ahead to proceed.

The next morning, at Boarshaw Boiler Works, I loaded into the Morris J4 van, oxy-propane cylinders, burning torch and gauges, the Pickhill electric-arc welding set, boxes of Rockwell Vertend welding electrodes and tool boxes holding a huge selection of chisels, hammers, drifts – and much more. Forty minutes later I arrived at World's End Piggeries. I was cordially greeted by Nuala O'Grady, who immediately proffered to bring a bowl of oat porridge and a steaming mug of tea into the Boiler House. I gave her the thumbs up, then Hughie, who had earlier received an instruction to assist me, helped me set up the burning gear, electric-arc welding set and the other tools and equipment at the rear of the Riley Brothers Scotch Marine boiler.

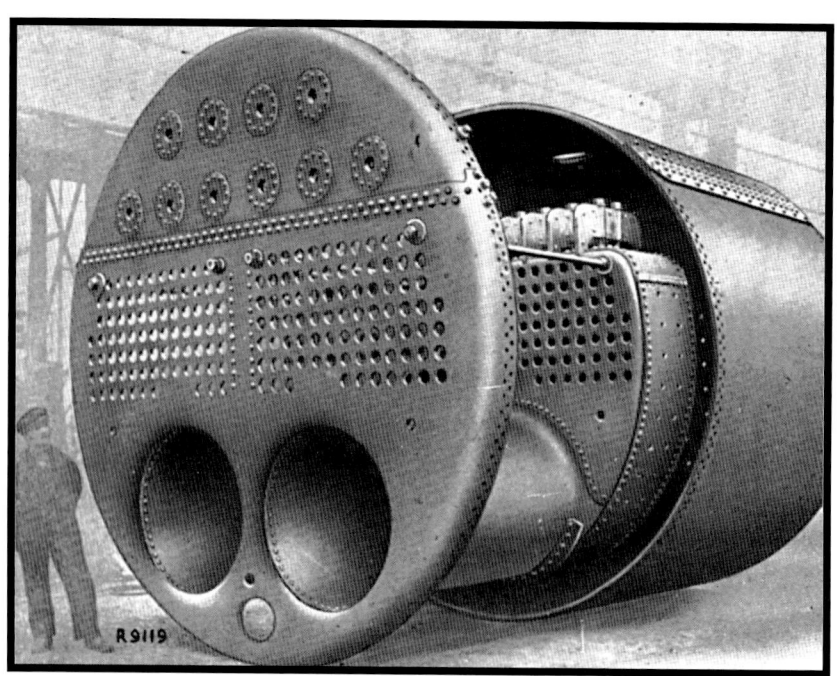

An exploded view of a twin-furnace Scotch Marine boiler showing the smoke-box tube plate, the furnaces, and the combustion-chamber.

My plan involved, primarily burning off the defective nuts, followed by carefully reducing the length of the projecting stay-bolts by burning, and then chiselling and peining the surrounding plate. Upon completion, ensconced upon a small metal drum, donned into my chrome leather welding jacket, my tweedy cap perched upon my head, onto which I firmly clamped my head shield, I then commenced operations. Thus, the first day sailed by pleasantly and satisfactorily. I was indeed most pleased with the welding work. I got on extremely well with ginger-nutted Hughie, who told me that he had been employed firstly as a pig herd, then after being collared by Nuala, his role changed to that of being her assistant in the maggot farm. He was, I considered, despite the age gap, clearly besotted with his extremely attractive female, Irish boss.

While I had been carrying out the welding work, Bart and Hughie, upon following my instructions had helpfully removed the Hopkinson twin-barrelled, spring loaded, safety valve, and had used the blank flange provided by myself to blank off the boiler.

By the late morning of my fourth day at World's End Piggeries, the welding of the stay-bolt ends was completed, and I then instructed Hughie to commence filling the boiler with water. Meanwhile, I was invited into the farm house dining room for a bowl of tasty, homemade Irish stew, which Nuala ladled out into a large deep bowl. There was also a plate of Lancashire muffins for dipping into the stew, and also the usual large mug of freshly brewed tea. Whilst heartily tucking in to this splendid repast, Nuala told me how Hughie and herself carried out the daily task of packing scores of wriggling white maggots into small, square, grease proof, paper-lined cardboard boxes. Around five o'clock, from Monday to Friday, the postman arrived in his little red Morris Oxford van to collect several sacks of the boxes which subsequently the GPO distributed all over Britain and Ireland; and also, to the farm's growing Scandinavian angling customers.

After lunch, I returned into the Boiler House, where Hughie brightly informed me that the boiler was now brim full of water.

Thus, with him perched upon the chequer-plate steel platform mounted on the boiler top, with his hand on the air-bleed valve wheel, I commenced slowly switching on and off the electric feed water pump. A mixture of aerated water was blown off through the bleed valve, resulting in us carrying out the first hydraulic test which amounted to 100 p.s.i. Because the boiler was generally steamed to a maximum of 50 p.s.i., this pressure being adequate for swill boiling, I therefore pressed the boiler to 100 p.s.i., twice the working pressure. With Hughie controlling the feed pump, I poked my head through the combustion-chamber man-hole to check for any leakage. There were just four small pin-prick leaks, which I peined to tightness with my hand hammer. The result was a totally water tight boiler repair. With Hughie's help, together we mounted the heavy refractory concrete-lined combustion-chamber man-hole door and nipped up the nuts.

Because the boiler was evidently uninsured, which was the usual practice carried out by the Lancashire Piggers in those far off days in the 1960s and early 1970s, therefore, there would be no visit from a Boiler Inspector. This was of no great concern to myself for I was 100 per cent confident in my workmanship.

Bart, upon returning into the Boiler House instructed Hughie to assist him in stacking wood and diesel-soaked rags onto the fire bars. With a whoosh, the fuel was quickly lit and when brightly burning Bart piled on several large shovels heaped with coal. Hughie kindly assisted me to load all of my tools and equipment into the Morris J4 van. I removed the coal-dust begrimed boiler suit and tweedy cap, shirt and string vest, then using my block of red carbolic soap to work up a lather, I enjoyed a strip wash in the Belfast sink located in a corner of the Boiler House. I had only just completed my ablutions, when Bart invited me to accompany him into the farm house kitchen.

"Me darlin' woife Nuala has baked us a foin ham and cabbage pie for tea. So ye'll join us?" he said smilingly.

"Hell's Teeth! World's End Piggeries reared ham! I didn't fancy any of that" I muttered to myself. I then thought, oh what the Hell, and followed Bart across the cobbled yard. Five minutes later, I was happily ensconced in the comfortable chair around the dining room table. Bart was lustily sipping neat Irish whiskey from a large glass tumbler. Raising the glass, he kindly offered me a tumbler of whiskey. I politely refused. Bart raised his glass, **"Alan, here's to you an' your fledglin' boiler repair firm. May the good Lord bring you much success."**

Leaning across the table, Bart handed me an envelope containing five £20 notes. Quickly counting the notes, I felt rather embarrassed, for the envelope contained far too much money. **"Alan, to be sure, Bart and myself are over the moon with your splendid boiler repair service. We can now start boiling swill right away. The money is a little bonus."** Said Nuala giving me a wide smile. I thanked them both for not only being much valued customers, but also for their kindness, hospitality and friendliness.

I thoroughly enjoyed the tea of ham and cabbage pie which tasted delicious. After shaking hands with Bart and Hughie and receiving a hug and a peck on the cheek from Nuala, feeling a trifle sad - for the last four days spent at O'Grady's World's End Piggeries had been most memorable - I drove out of the cobbled yard. Glancing out of the driver's door window I discerned great volumes of steam rising above the roof of the Swill House. I felt immensely satisfied.

Bart O'Grady's World's End Porkers.

Chapter 5

THE BLACK BIRD CHOCOLATE WORKS JOB.

One fine, sunny yet bitingly cold winter's morning, I was inside my small Boarshaw Boiler Works with the large wooden doors swung wide open letting in the cold air. Feeling content with the world and myself, using a 3 lb hand hammer and a short 'crocodile-gob' knocking out drift, I was happily engaged belting out over 60, one inch copper tubes out of the tube plates of a 5 feet long, carbon steel shell of an antiquated calorifier vessel.

During the previous afternoon, I had received a telephone call from the Chief Engineer of a Salford confectionery works, who breezily informed me that his firm's insurance company Boiler Inspector had commended my name to him. The Chief went into some depth of details regarding the 'steam to hot water' generating calorifier, evidently which was the only one of its kind his firm had. Therefore, there was an urgent requirement to carry out the full re-tube as per the instructions of the Boiler Inspector. He also mentioned that his own fitters were in the process of dismantling the connecting steam and water pipe work and upon completion the vessel would be awaiting collection in the chocolate firm's own loading bay.

"Golly Gosh!" I muttered out aloud. This job being of an industrial flavour was indeed just up my street. With rising excitement, I almost bellowed down the telephone receiver that I would be most pleased to visit and survey the calorifier. The Chief then gave me the verbal green light to proceed followed by an Order Number, which would be required on the invoice.

Thus, the telephone call left me feeling ecstatic, for due to the recent jobs carried out on Piggeries, I was desperate to escape the rich pig muck pong that appeared to cling onto all of my tools, burning gear and welding equipment; and also, the Morris J4 van. Within less than a half hour, I was driving the van through Middleton town centre, where upon taking Manchester Old Road, soon I passed through Rhodes village, then Blackley, and on through busy Cheetham Hill which led me via a bewildering maze of side roads, until, Liverpool Road, the A6 was reached.

While trundling through the heavy traffic near Manchester Docks, I suddenly spied one of Phoenix Boiler Maker's, large Ford Thames Trader drop-side lorries which was parked in front of the huge oil refinery at Weaste. Hitting the brakes, I slowed the van down and sounded the horn to alert the two men who were unloading long, steel boiler tubes from the back of the lorry. Upon hearing the sound of the horn, a brawny chap sporting a crew cut hair style looked up and upon recognising my name displayed on the side of the van raised his hand in acknowledgement, smiled and shouted a greeting.

It was Tommy, a member of Carrot's Squad who lived in Blue Pits. I waved back. The utmost feeling of pride rose up into my chest, for here I was, a young 21 year old lad, striking out into the World of Boiler Making; and was this morning actually travelling to collect an interesting and hopefully highly profitable industrial job. Due to the heavy flow of Dock's traffic, I continued my journey. I was well acquainted with Tommy for throughout Phoenix Boiler Makers he was known as the 'bread van driver basher'. *(See RIVET LAD – Lusty Tales of Boiler Making in the Lancashire Mill Towns of the 1960s).*

I knew that upon Tommy returning into Phoenix Boiler Works, he would spread the word around that young RIVET LAD, Alan McEwen appeared to be successfully getting loads of industrial boiler repair work. And making pots of cash!

Driving along doing my best to avoid being hit by one of the huge numbers of heavily laden lorries that thundered by, stealing a glance down at the chocolate work's written address details stuck onto the van's metal dashboard with a small magnet, I realised that I had almost reached the locality of the confectionery works. Following the written directions, I turned the van left to enter a cobbled street lined with long rows of red brick houses with ground floor bay windows. I noticed these windows were hung with lace curtains probably to stop nosey parker passers by from looking inside. Most Lancashire cotton town house windows were draped thus.

This cobbled street was considerably quieter and traffic free than the bustling A6. As I trundled over the cobbles crawling slowly, I noticed three young lasses happily playing hop scotch on the flag stone pavements. As I passed they waved. Reaching the end of the street, I spied a huge set of iron gates set into a high wall of smoke-blackened brick. The name of the confectionery firm was emblazoned upon a large board fastened to the wall: BLACK BIRD CHOCOLATE & CONFECTIONERY CO. LTD. One of the gates was closed, but the other gate was set back to allow access for lorries and vans entering the Works. I slowly drove straight in. The first thing to hit me was an absolutely, enchanting aroma of boiling sugar and the delicious perfume of hot chocolate that permeated the factory surroundings and the streets beyond.

The appearance of the smoke-blackened surrounding buildings identified that the factory was a former old cotton mill. I continued driving across the extensive tarmacadamed Work's yard. Upon noticing a tall, palatial-looking, narrow brick building with a huge bow-topped window, due to the architectural detailing and fine stone mouldings I knew this building was the Engine House, which originally powered the former cotton mill. I just wondered with rising interest if the mill steam engine remained within.

Situated adjacently, was a typical Boiler House, or fire-hole – as these distinctive buildings originally housing one, two or more coal-fired Lancashire boilers were known and commonplace throughout Lancashire. Further still, I could discern the unmistakable whirring sound of a lathe probably carrying out a turning operation on a steel shaft. In the background also, was the clanging sounds of a hammer striking upon an anvil: both sounds identified the firm's Engineering Work Shops. Within I guessed, would be the Chief Engineer's office.

I parked the Morris J4, and upon taking a quick glance at my appearance in the wing mirror, induced me to do up the top buttons of my boiler suit which was clean on that morning. My mop of white-blonde hair was well slicked down with Vitalis hair oil. I stepped into the doorway and looked inside. The Work Shop machinery, in spite of looking quite elderly, appeared to be nevertheless, well maintained and painted light green. There was a huge Asquith of Halifax radial-arm drilling machine, a large Dean, Smith & Grace of Keighley centre lathe, a smaller turret lathe of a make I didn't recognise, two vertical Herbert pillar drills, and a Rapid mechanical hack saw. My eyes also took in, a well-equipped Blacksmith's forge and a huge steel rack containing countless tongs and other metal forging tools. Tucked away in a corner was a large steam powered, Scottish-built power hammer. There was also a welding booth containing a large oil-cooled electric arc-welding set, a row of steel benches, complete with massive engineers' vices, a large square steel, fabrication and welding bench and several large, metal 'pigeon holes' holding a variety of boxed welding electrodes. I could also make out a set of oxy-acetylene gas welding cylinders, pipes and gauges.

As I slowly walked across the floor of Staffordshire Blue engineering bricks, one of the fitters, a rotund, balding chap, looked up from his bench.

"Hiya cock, you'll be lookin' fer' th' Chief Engineer, Donald Ferguson?" he inquired. I nodded. *"Can'st see yon door?"* Jabbing an oil-stained hand towards what appeared to be an office door. *"That's th' Engineer's office, an' he's in. He'll be suppin' a brew o' tea about this time."*

I gave the friendly fitter a smile and tramped across the oil-stained brick surface of the floor, swerved around a gigantic, globe-like mass of swarf that blocked my way, and upon reaching the office door, I gave it a sharp rap with my balled fist. *"Who's there?"* The door was then thrown back to reveal the figure of a short, stout man of middle years. He wore round brass-rimmed spectacles with thick lenses which were perched on the bridge of his wide, flat nose. The Chief Engineer wore a white boiler suit which was at least two sizes too small, with several missing buttons. I noticed his hands were quite small and bony. One of which was grasping the handle of a mug of tea.

I quickly introduced myself stating I had come to collect the calorifier requiring re-tubing. In fact, I had addressed Mr. Ferguson as Chief; adding sir.

"Aye, good mornin' young mon. Me name's Donald. Here at Black Bird Chocolate we don't like bein' formal. We're just hard graftin' Salfordian engineers. And I'm just one of th' lads. What did ye say yer name were?"

I felt a little taken aback – for during my apprenticeship at Phoenix, it had always been drummed into us young chaps, that all managerial staff, and particularly Boiler Inspectors and Chief Engineers must always be formally addressed. We shook hands. Donald shouted an instruction to one of his engineering staff who had just re-entered the Work Shops.

"Larry, mash a brew cock, for th' boiler maker wilt tha?" Within a couple of minutes, a steaming mug of tea arrived together with a cracked, grubby, plate containing an array of silver-paper wrapped bars of dark and milk chocolate.

"Thee get suppin' an' grab a couple o' bars o' our very own chocolate, Alan. Aye, this bonny tastin' chocolate is made right here in our factory."

I was rather partial to milk chocolate, so I helped myself to a bar. Removing the silver wrapper, I bit off a huge chunk which was washed down with the tea. It felt really good and refreshing.

"Has't finished suppin'?" Questioned Donald. I nodded and with my jaws working on the remainder of the chocolate bar, I followed the Chief Engineer out of the Engineering Work Shops the short distance to the adjacent building which was the Boiler House. Within were a beautiful presented pair of Lancashire boilers, which I discerned from seeing the four large oil burners, had been converted from coal-firing to heavy oil. Unusually, the brass D-Rings remained in place bolted around each of the furnace mouths, proudly proclaiming the boiler maker's name, date and the work's numbers: Foster, Yates & Thom. 1932. These D-Rings were highly polished. Generally, when converting from coal to oil-firing the D-Rings would be removed. The boiler's

front-end plates and boiler shells were heavily lagged in asbestos; the steam and feed water pipe work were also covered in thick asbestos lagging, all of which was sheet metal-cladded and painted bright red.

A pair of heavy oil-fired Lancashire boilers.

I noticed that each boiler was fitted with a set of highly polished, bronze water gauge cocks, a bronze syphon and cock surmounted with an 11-inch diameter brass pressure gauge displaying the working pressure of 160 p.s.i.

"Ah see, tha's impressed wi' me bonny owd Lankys. Both are in fine fettle." Donald then informed me that both boilers had been inspected in August of the previous year. *"The Scotch man – as we call him, Shafto McKenzie, th' Boiler Inspector just asked fer't fusible plugs be removed, an' th'oles plugged off. This was due to both boilers havin' been converted to run off heavy oil, a year earlier."*

I was indeed, most impressed with the beautifully clean, neat and tidy Boiler House and particularly with these attractively turned out Lancashire boilers. While I was gazing around the boiler plant, swimming before my mind's eye was the appearance of a friendly, heavily wrinkled face and the laughing eyes of Mr. Shafto McKenzie, C.ENG. M.I.MECH.E.; M.I.MAR.E. I quickly realised this was the genial Scottish Boiler Inspector I had previously met while working with Carrot Crampthorn's Squad at McClintoch's Mountain Mint Rock & Toffee Works, at Primrose Bank, near Denton, where I had assisted with the re-tubing of a Thomas Beeley & Son-built Economic boiler. *(See RIVET LAD – Lusty Tales of Boiler Making in the Lancashire Mill Towns of the 1960s).* Mr. McKenzie must have indeed been really impressed with myself and my boiler making skills, for it was definitely that gentleman that passed my details onto Donald Ferguson. Casting my eyes over the Foster, Yates & Thom Lancashire boilers, I felt deeply honoured to be the chosen one to carry out the re-tube of the calorifier.

After I backed the Morris J4 van into the Loading Bay, two of Donald's fitters, Larry and Willie, by operating a set of chain blocks suspended from a steel girder, slung the timber crate containing the calorifier onto a steel roller which I had placed onto the van's back end.

As Willie slowly lowered the calorifier crate onto the roller, Larry and myself commenced pushing the load further into the van, until, with much huffing and puffing, we achieved our goal, the load was safely stowed. I gave the pair of ever-helpful, friendly Salford fitters a friendly wave as I drove the heavily-laden Morris out of the Loading Bay towards the factory gates.

I had given Chief Engineer Donald Ferguson, a cast-iron guarantee, that I would return the completely re-tubed calorifier within 7 days. This was indeed a tall order, but as the Trafford Park tube specialists had plenty in stock, they had promised me that I could immediately collect from their premises the full bundle of copper tubes.

Arriving back at Boarshaw Boiler Works, I would now have to overcome the rather difficult job of unloading the calorifier from the Morris, without any harm coming to the vessel and without any form of lifting equipment whatsoever.

However, my blossoming, friendly relationship with Alec, the coal-yard owner would be most fruitful, for I knew he had a powerful tractor which had a front, loading bucket. Alec certainly obliged. He drove the tractor up to the open rear doors of the Morris van, and after a spot of gentle pulling and careful lifting of the rope slings he had kindly loaned me, which I had fitted around the calorifier, the vessel was successfully lifted out of the van. Upon my instructions, Alec lowered the calorifier onto a pair of low, steel trestles, which I had recently fabricated from second-hand, rusty channel and angle irons. I had previously purchased this old iron off a local farmer for a trifling sum.

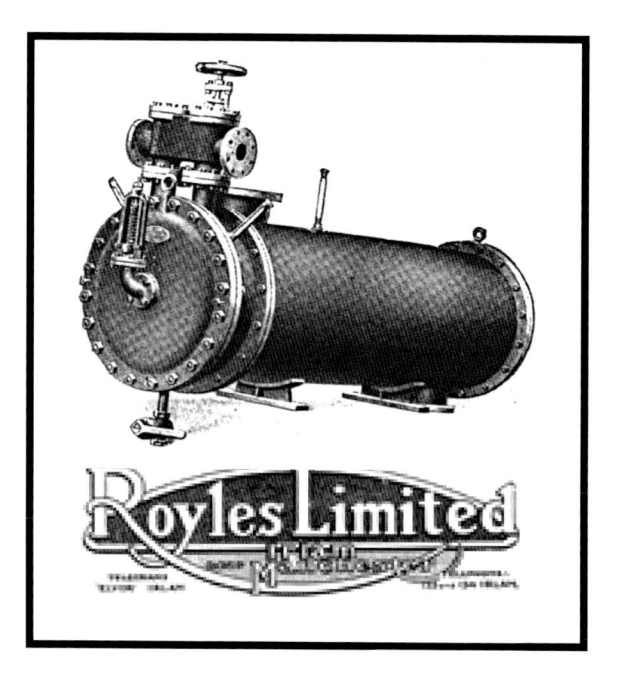

Royles Limited tube and shell-type calorifiers.

With the calorifier vessel safely mounted upon the trestles, I got to work by firstly, applying a diesel oil-soaked brush to lubricate the nuts and bolts securing the dished, cast-iron end-plates. While allowing the diesel oil to soak into the threads, meanwhile, I mashed a brew of Horniman's tea.

A couple of months previously, I had set up several 12-inch diameter cast-iron pulley wheels, together with ropes, which were bolted onto the Boiler Work's angle iron roof trusses. I operated these rope pulleys for lifting heavy objects onto my welding bench. The maximum weight I had previously lifted was around one hundredweight. The cast-iron dished end-plates would perhaps weigh a trifle more. Thus, after clamping a lifting D-shackle onto the connecting ring, I commenced hauling on the lifting rope resulting in the lift taking place easily and safely. I lowered the end-plate onto the floor, then commenced the operation with the other cast-iron dished end-plate. I then skidded both out of the way.

My plan was to quickly hammer out about six copper tubes. This would enable me to carefully check the diameter and gauge, which was crucial prior to ordering and collecting the replacement tube bundle from Trafford Park. Fortunately, all went well, and after jotting down the length, diameter and number of tubes required, I telephoned the order through to the tube supplier who again invited me to collect the tubes at my discretion. Thus, I immediately drove the Morris J4 van down to the long-established heat exchanger and calorifier tube supply firm in Trafford Park.

Because I had placed the order via the telephone, within about 10 minutes of my arrival at the offices, a friendly salesman, led me into a vast warehouse, where I spied a plywood packing case sporting a printed label stuck on top, which gave the name of my business: H.A. McEwen (Boiler Repairs). Boarshaw Boiler Works, Middleton. Upon checking the transaction note handed to me by the salesman, all appeared in order. Back in the sales office, I handed over a weighty bundle of £10 notes in exchange for an invoice stamped: PAID IN FULL.

Upon my return to Boarshaw Boiler Works, without further ado, I recommenced using the hand hammer and crocodile-gob drift, the task of driving out the condemned copper tubes. The job progressed extremely well, and by tea time all of the tubes were out and lying on the floor. I then used the hosepipe connected into the water supply tank up on the roof trusses, to deliver a jet of water to wash out the dirt and loose scale from within the calorifier shell. This left the vessel and tube plates clean. By using a sanding disc fitted into my Black & Decker electric drill, I sanded off the surface of the tube plates. I then carried out a close visual examination of the tube holes and the inside of the vessel which revealed all was satisfactory.

A calorifier with copper tubes undergoing re-tubing.

I broke open the plywood packing case lid containing the batch of gleaming brand-new copper tubes. Picking up one of the tubes I threaded it through a bottom hole in the nearest tube plate. It entered easily, and upon pushing further entered the hole in the opposite tube plate. Phew! That was easy, I thought. Within about 10 minutes all 64 of the one-inch copper tubes were in place between the calorifier's tube plates.

Using four light screw-tight clamps – one of several I had made while working at Phoenix Boiler Makers – I firmly clamped together 8 tube ends. I then selected a 5 roller Wicksteed tube expander and mandrel and adding a squirt of light oil from an oil can into one of the bottom rows of tubes, I inserted the tube expander and mandrel. Placing the ¾ inch square ratchet onto the square end of the mandrel, I then commenced the rapid-revolving action which expanded the tube ends tightly into the tube holes. Thus, one after the other the tubes were expanded.

By using several blank flanges, borrowed from Black Bird Chocolate Work's Engineering Workshops, and a quantity of nuts and bolts from a hessian bag of fasteners I had been accumulating from well before I left employment at Phoenix, thus, I soon had the calorifier's flanged connections blanked off for the forthcoming hydraulic test. Using the hose pipe from the water storage tank, I began filling the calorifier. Keeping a wary eye on both tube plates, I continually checked for leakage. Thankfully, the tubes remained tight. But would they leak under hydraulic pressure?

Using the hydrostatic pressure, and operating the ¼ inch brass valve screwed into the topmost blank, I bled the air out of the calorifier. I then closed the valve, and commenced applying pressure via the small test pump, which I had recently purchased together with several ex-military tools from an army, navy and air force surplus store in Haworth, West Riding of Yorkshire. This occurred during a fleeting visit to the fledgling Keighley & Worth

Valley Railway's locomotive shed, where I had carried out a minor welding repair on the boiler of a L&Y.R. tank engine.

Upon applying the full hydraulic test pressure of 50 p.s.i. - (the maximum working pressure of the calorifier was 25 p.s.i.) – a few tiny drips were revealed. However, these were soon cured by a few nips with the tube expanders. Thus, I was entirely satisfied with the work. The whole job had only expended 4 working days. Therefore, I would easily keep my promise to the Chief Engineer. Late the same afternoon, I telephoned Donald Ferguson informing him that the calorifier was now fully re-tubed and ready for returning. Donald was absolutely cock-a-hoop. I promised him I would deliver the calorifier, together with all the scrap copper tubes the next morning.

Then, back into Alec's coal yard I went begging the loan of the tractor. Within half an hour, the newly re-tubed calorifier had been safely loaded into the Morris J4, together with the bundle of scrap copper tubes. The poor Morris J4 was so severely over-loaded that it looked to be sat upon the road surface!

Next morning at 8.30, my van was safely parked inside the Chocolate Work's loading bay, where Donald's fitters assisted me with the unloading. One of the fitters, Willie, brightly informed me that the Boiler Inspector, Shafto McKenzie was visiting that afternoon to witness the second hydraulic test of the calorifier which he and his mate would carry out. Thus, my job was finished.

After enjoying a huge mug of tea and helping myself to a Black Bird Chocolate Bar, I bade the delighted Chief Engineer Donald Ferguson and his men a hearty cheerio. I then fired up the Morris J4 van. A shout rent the air, **"Whoa Alan. We have a few bars of our delicious Black Bird Chocolate for you. And let me say how pleased I am with your excellent service."** Exhorted Donald with a huge smile.

Upon arriving back into Boarshaw Road, I spied a huge, 10-ton tar-melter, undoubtedly requiring the urgent replacement of the burnt-out pan bottom parked slap bang in front of the Boiler Works doors. Hell's Teeth! Another blessed rush job from Asphalt George!

A typical 1960s-built tar-melter.

Chapter 6

OLD GOLDEN PITS PIGGERY JOB.

On a morning of dark low cloud and sweeping rain, I was happily working within the warm, dry comfort of my Boarshaw Boiler Works, fabricating a steam manifold from a five feet length of six inch seamless carbon steel tube with domed steel end-caps, and a variety of flanged stools. I was indeed glad to be out of the heavy rain and working on this interesting and very profitable pipe-fab job for my recently acquired new customer, the Sundance Chemical Works in Newton Heath.

Suddenly, the telephone in my tiny office rang out. Grabbing a lump of cotton waste and wrapping it around the receiver to avoid it becoming smothered in the soot and oil concoction that covered my hands, I placed the instrument up against my right ear. The croaky-rasping voice of a 70 Capstan Full Strength per day smoker thundered out of the receiver. **"Hello, hello. Is that th'boiler chap? Hello theer, I am a pig-breeder, tha knows, a Pigger frae Owd Golden Piggery. Me name's Jake Belthorn."**

Smokey Jake, then explained between prolonged coughing fits, that he occasionally called into the Gardener's Arms, one of the two pubs in my home village of Top of Hebers. Evidently, while recently enjoying a couple of pints, he had bumped into one of our local characters, an elderly, retired pig-breeder, Percy Ramsgill.

Both old Piggers chatted on full throttle about pig stuff: breeding, fattening and then about the collection and transportation of pig swill. They moved the talk onto the merits of good steaming boilers, pig swill boiling and the high price of corn and other manufactured pig foods. Both agreed, swill was the most nutritious. And cheapest.

After supping four pints of mild ale, and smoking eight tabs, Jake bade local chap Percy a good night. Then after another huge coughing and retching session, he walked out of the public bar door. He coughed up a huge ball of phlegm, which he spat onto the pub's cobbled yard. He staggered, his belly full of strong beer, which was making him feel a little light headed. Upon reaching his old, much abused Bedford pick-up lorry parked in the dark around the back of the pub, Jake jammed himself into the gap between the lorry and the stone wall, he then emptied his full bladder, while sparking up another Capstan Full Strength tab.

Evidently, upon him mentioning to Percy Ramsgill that he was looking to sell an old Vertical Cross-tube boiler and a steam jacketed pan - ideal for boiling pig swill, Percy, whom was well acquainted with myself, then handed Jake one of my recently printed business cards urging Jake to phone me. Jake had been jabbering down the phone for what by now seemed like hours. The old Pigger spoke non-stop like a machine gun. I patiently listened, and when he paused to cough and retch, I then inquired how I could assist him.

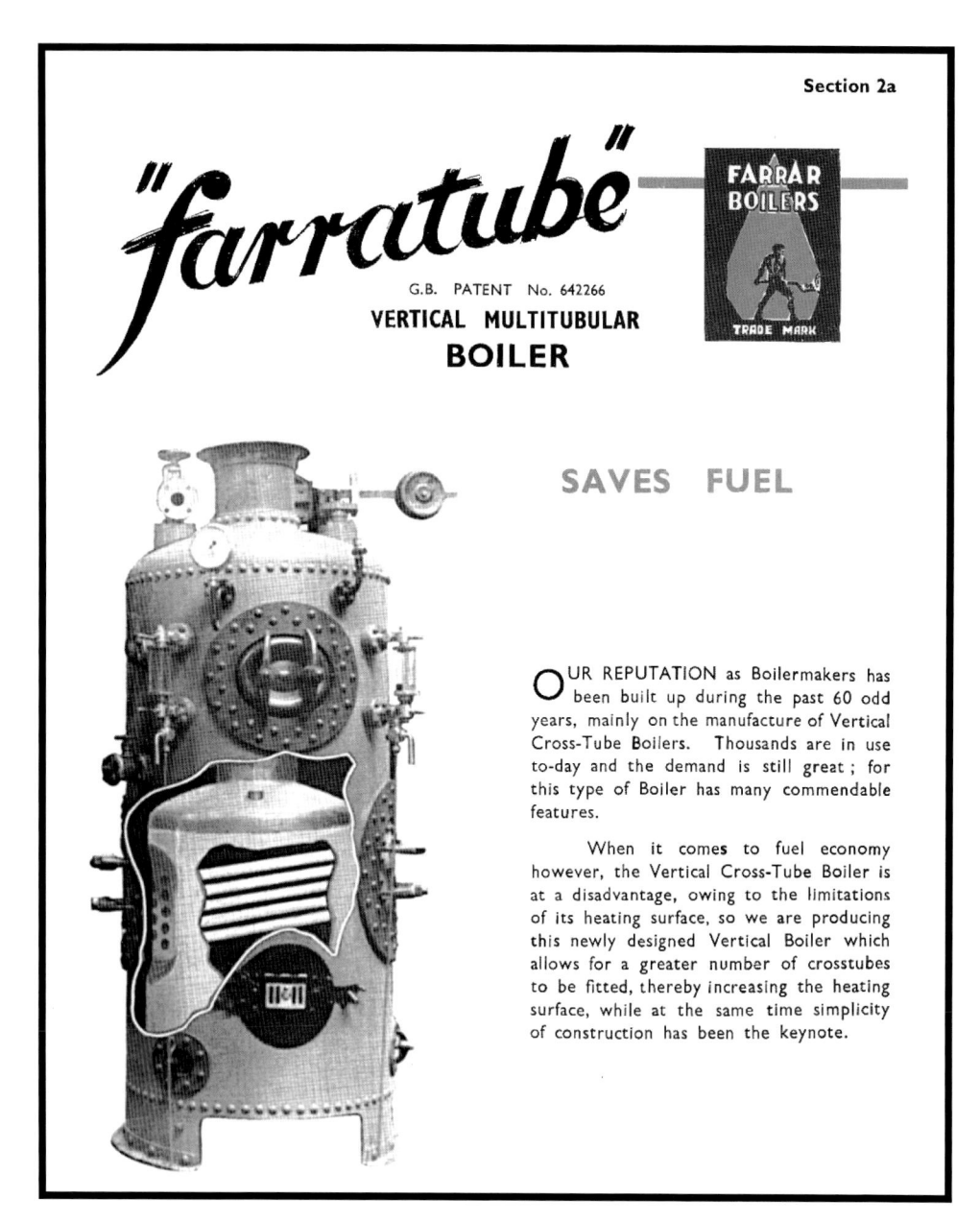

"farratube"

Section 2a

G.B. PATENT No. 642266
VERTICAL MULTITUBULAR BOILER

FARRAR BOILERS
TRADE MARK

SAVES FUEL

OUR REPUTATION as Boilermakers has been built up during the past 60 odd years, mainly on the manufacture of Vertical Cross-Tube Boilers. Thousands are in use to-day and the demand is still great; for this type of Boiler has many commendable features.

When it comes to fuel economy however, the Vertical Cross-Tube Boiler is at a disadvantage, owing to the limitations of its heating surface, so we are producing this newly designed Vertical Boiler which allows for a greater number of crosstubes to be fitted, thereby increasing the heating surface, while at the same time simplicity of construction has been the keynote.

Farrar Boilers Ltd, The "Farratube" Vertical Multi-tubular Boiler.

Jake explained between coughing fits that he was the owner of a Vertical Cross-tube boiler and a riveted steam jacketed pan he had used for many years for swill boiling. He believed both items were in excellent order and he wanted to sell both. If they were of interest to myself, then what's the best I could offer?

After another long-winded chat, I informed Jake I could actually visit his Piggery that very afternoon. He agreed and gave me directions. With Jake's rasping voice banging like a drum inside my head, I then switched on the electric kettle to mash a mug of tasty Horniman's tea, and from my snap bag I produced one of my mother's Lancashire cheese sandwiches.

Driving the Morris J4 van along the Rochdale to Oldham Road, I saw a road-side sign: Black Pits Clough, beneath which was a smaller much weathered large stone block onto which was carved Old Golden Pits. The signs pointed the way up a steeply inclined, rough cobbled stone track that worked its way up to the hill top. My poor old Morris's engine grunted and growled as I drove the van up the vertiginous, cobbled track. Either side was dense hawthorn thickets growing amongst piles of yellow brick rubble.

Upon reaching the crown of the hill, I then noticed in front of me a large yard, and another sign proclaiming Old Golden Pits Piggeries. Thus, after my spot of mountain climbing, I had arrived.

Old Golden Piggeries appeared to comprise of a clutch of curious-looking, ramshackle barns, sheds and former pig sties that appeared to be built from roughly-mortared, yellowish fire bricks, old colliery pit props, long lengths of ex-railway crossing sleepers, sections of rusty, riveted steel plate and with roofs of rust-eaten corrugated iron. The whole yard and buildings had encircling walls constructed from the same yellow-hued fire bricks. I parked the Morris up against one of the walls and upon clambering out quickly realised that the yard surface was several inches deep in liquid, yellowish clay that almost covered my boot tops. I will endeavour to tread carefully hereabouts, I thought.

Looking at the wall I noticed that many of the bricks appeared to be grossly misshapen, and of a wide variety of sizes. The surrounding buildings also looked like they had been constructed from the same curious-looking yellow bricks. What a strange looking location I am visiting, I thought.

Moreover, despite these premises being a Piggery, I couldn't discern the usual sounds of squealing, grunting porkers. Nor was there any nose-curling, offensive aromas of boiling swill or worse; rotting pig muck to pervade the surrounding moorland air. Instead, my nostrils detected the pleasant perfume of heather wafting down from the nearby moors. There were also no sights or sounds of human activity!

Looking about me, I could see that Old Golden Pits Piggeries was located among what appeared to be a series of old clay pits, surrounded by large mounds of yellowish clay waste and gravels. There were also a number of deep water-filled pits.

With my fascination for studying industrial remains by now well risen, I ventured into the disused clay workings. Due to the recent rain the ground underfoot was even more slippery than in the yard. My heart leapt a bundle upon noticing rusty, narrow gauge, iron tram tracks embedded in the clay and gravelly ground that sinuously ran all over the extensive clay pit site.

Intrigued, I followed them uphill to where they terminated in a small yard, whose floor was constructed from the ubiquitous yellow brick. Scattered all about were the rusty, extremely weathered remains of several, four-wheeled, clay tubs, which, I guessed had been once used for transportation of the mined yellow clay and gravels.

A small, roofless building lay about a further ten yards ahead. I became extremely excited. Perhaps inside the building there could be a small, narrow gauge locomotive. Upon reaching the woebegone-looking building however, I was disappointed for there was no tram locomotive within.

There was however, the fascinating remains of what appeared to be a steam-powered marine winch, with a pair of vertical inverted steam cylinders bolted either side of a robust cast-iron pillar. Upon noticing a name cast into the pillar base, I picked up a stick to scrape away the thick, cloying layers of oil and muck. A handful of grass took the place of a rag. Thus, with the oil and muck wiped away, the steam-winch maker's name was revealed: THOMAS LARMOUTH & COMPANY. MAKERS. TODLEBEN IRONWORKS, SALFORD.

Generally, the winch excepting for the main steam valve and other minor brass fittings remained intact.

Looking down on the sticky, rain-soaked yellow clay ground revealed several, heavily rust-eaten pieces of what must have been the original steam supply pipe. My guess was that the steam-winch had originally been supplied with steam by a 2-inch iron pipeline running up the yard from the Boiler House.

I plonked myself down on an upturned oil drum to drink in the scene of the dis-used yellow clay and gravel mine. My fertile imagination conjured up the miners of 100 years earlier, hard at work, digging by spade and clay-mattocks, lumps of the yellow, sticky clay which was then loaded into small riveted, iron plate constructed, 4 wheeled tubs mounted on the tram rails. The miners themselves, must have been a rough and tough bunch of hard working men, who endured the harsh conditions of clay mining in all weathers, would load the tubs by hand.

When full, the men would physically push the tubs from all areas of the mine to the small yard where the Thomas Larmouth steam-winch was located. Here, the tubs would be coupled together and then lowered by the steam-winch down the hill-side and into the Main Yard. From there, the raw clay materials and gravels would be stored inside the huge brick barns awaiting to be sold and then dispatched to brick-making and construction firms all over Lancashire and beyond.

I was in my very own world, and enjoying my thoughts. Suddenly, I was brought back into the real world by the sound of a man shouting. *"Hello, hello Boilerman, I've seen your van, but where are you. Tha's bloody trespassin'."* Then a tall, walking-stick thin man in his late sixties appeared.

I arose from the oil drum, and walked towards the elderly man who wore a much-stained and tattered tweed jacket. He had a bald, monk-like pate, and had long, white greasy hair that hung down to his shoulders. Perched between his lips was a glowing cigarette stub. I proffered my hand as I uttered a friendly greeting and introduced myself. The old man did not reciprocate. *"Now young fella, tha's bin pokin' about on me land and around me buildin's. I do hope tha weren't up to no good like nickin' scrap brass?"*

I assured him, that upon my arrival at Old Golden Pits Piggeries, and with there being no-one to meet me – for he was late by around 30 minutes – this was the reason, coupled with my deep interest in old industrial sites, that I had been exploring. He then broke into a smile, the cigarette stuck with saliva onto his lower lip, remained lit. *"Reet lad, aye theers nae bother. I con see that th'art a decent young fella. I get a bit over cautious now and again, for th' clay pits are quite remote, an' I've had a large amount o' scrap metal pilfered."*

He then grasped my hand. *"Welcome to Owd Golden Pits. Reet, foller me down to th'office, an' I'll mek us a brew. I'll then show thee into th'boiler house, wheer tha'll see my reet bonny, Vertical Cross-tube boiler, an' me owd pig swill boilin' pan."* We then walked down the track, the wet and sticky clay clarted my boot soles.

The clay mine office measured about 12 feet by 10 feet. Stacked around the tall book shelves and open-fronted cupboards lining three walls, were huge stacks of mouldering day books, ledgers, wages books and hundreds of leather-bound files. Everything was deeply covered in yellow dust. There was a small window that shared the same wall as the door whose glass panes were thickly blanketed in the ubiquitous yellow dust; into which countless bodies of dead flies and insects were embedded. Hanging from a chain suspended from the centre of the dark brown painted ceiling, a single electric light bulb delivered a small pool of light over a huge old oak desk, whose leather top was badly cracked with sections worn through to the wood.

The elderly Pigger, Jake, filled the ancient electric kettle with water from a heavily patinated brass tap fastened onto an iron bracket protruding out of the wall. Beneath was a filth-encrusted Belfast sink. He swished out the stubborn yellow clay dust from a pair of chipped old pottery mugs by inserting two heavily tobacco-stained fingers. With a banshee-like shriek, the kettle reached boiling point. Jake poured the boiling water onto the tea in the mugs. **"Bugger me cock, we've getten nae milk. Tha'll hev to sup thee tea black. But we've sugar."** He said, smiling through his disgustingly broken and tobacco-stained false teeth. Using a fore finger, he dug into the sugar bowl, scraped up a number of dead insect bodies, and neatly flicked them onto the floor. He then dug a tarnished old spoon into the congealed sugar and dropped a heaped spoonful into each of the hot steaming mugs of tea.

While waiting for the scalding liquid to cool, I asked Jake what he could tell me regarding the history of the Old Golden Pits clay mine. Lighting up another Capstan Full Strength tab, and sucking down lungfuls of smoke, the old Pigger started to relate how the history of the clay mine workings evidently went back to around 1860 when two local brothers named Whittle, expecting to locate coal, commenced digging into the hillside. There was numerous coal 'day pits' in the locality. Unfortunately, the brothers found no coal. What they actually discovered, the yellow fire clay and gravels, would overtime turn out as equally profitable as the black stuff. The Whittles soon found an eager buyer for their yellow clay: a fire brick and refractory manufacturer based in nearby Oldham, who subsequently purchased every cartload of clay won from the mine, and for a great many years thereafter.

Eventually Whittle Brothers installed several steam-powered haulage winches, purchased a large quantity of wheeled iron tubs for transporting the mined clay around the site; constructed a tram road hauled by the Thomas Larmouth steam-winch – (whose remains I had earlier viewed) – built a Boiler House which contained a coal-fired Cornish boiler, and generally established what was to become an extremely profitable concern.

Following the Second World War, in the early 1950s, the mine's deposits of easily won, quality fire clay were rapidly becoming worked out. By 1960, the mine closed down. The Cornish boiler now in a poor condition due to neglect and bad feed water, was cut up on site for scrap. Likewise, much of the other plant and equipment scattered about the clay mine, including the steam-powered tram haulage system. The only winch to survive was the example I saw. Both the Whittle Brothers died in 1962, and the whole clay mine was put up for sale. Mainly owing to the prevailing bad business conditions throughout the 1950s and into the early 1960s, Old Golden Pits, clay mine remained unsold.

In 1964, Jake Belthorn, the younger brother to a Rochdale pig farmer, having received a cash legacy from a deceased aunt, purchased Old Golden Pits, lock stock and barrel. By the Christmas of that year, Jake had set up his own Piggery of 400, high quality porkers. At a closed-down factory auction, he purchased the second-hand Vertical Cross-tube boiler and the steam jacketed pan. After installing the VCT into the old Boiler House, and the jacketed pan into an old farm cottage which he later called 'Swill Cottage', both the steam plant and the pig fattening business served him well. Unfortunately, however, over a period of 12 months he had increasingly felt unwell with continual painful chest pains and a racking cough. He found he could no longer work, and therefore sold off his herd of quality pigs and pocketed a princely cash sum.

Thus, Jake had managed to sell off all of the valuable stock and equipment excepting for the VCT boiler and the steam jacketed pan. I followed Jake across the slimy surface of the yard. The old man had a glowing Capstan Full Strength tab clamped between his lips. As we approached the Boiler House, which was similarly constructed as the other buildings of roughly moulded yellow fire bricks, the elderly Pigger faltered, and then commenced to violently cough. He spat out the tab, and upon leaning over the rust-eaten remains of an old iron water tank, the sick old chap vomited. Wiping his mouth with the sleeve of his jacket, and clutching his chest, he quietly muttered, **"Tis the bloody cancer. The doc's given me less than six months to live."** He then fished another cancer stick out of a packet and sparked it into life with a Swan Vestas.

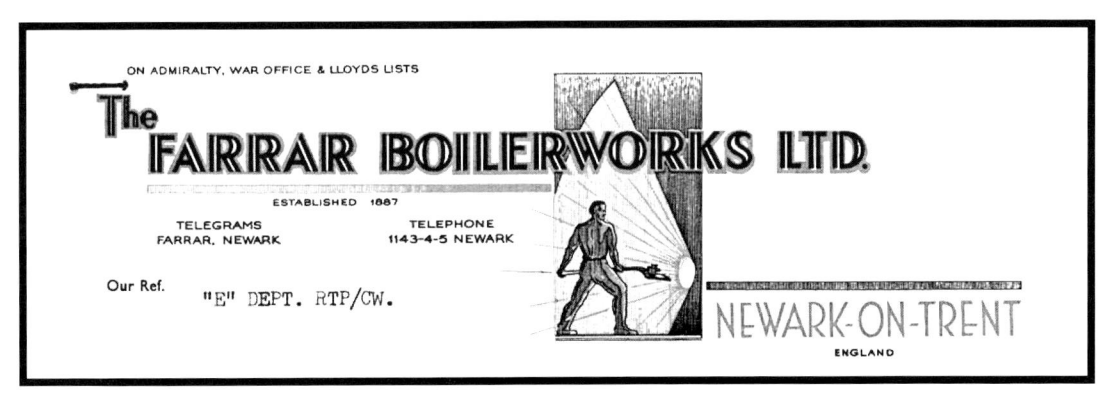

The Farrar Boilerworks Ltd. Letterhead.

The Vertical Cross-tube boiler built in 1959 by the Farrar Boiler Works in Newark-on-Trent stood about 12 feet tall with a diameter of around 5 feet. Excepting for it being covered in fine yellow dust, it appeared to be in good condition. A wooden ladder was leant up against the boiler, providing access to the man-hole. The man-lid had previously been removed, and was precariously perched up on the domed top. Walking around the boiler, I noticed that two of the mud-lids around the foundation ring had also been removed. This afforded reasonable access to look into the water side of the fire-box, the foundation ring, the uptake and cross-tubes.

I motioned Jake to sit down on an old chair and take it easy. Using my slim battery-powered torch, I shone the light into both of the open mud-holes. Except for a minor amount of dust and loose scale, there was no evidence of metal wastage or corrosion. Up the shaky timber ladder, I climbed. Shining the torch through the open man-hole revealed the fire-box crown to be scale and corrosion free. However, on the uptake tube, there were a small number of 'wasted' areas and one large 'pit'.

Nevertheless, as I could easily reclaim the wasted metal areas by applying the electric arc welding 'buttering' procedure, I was therefore generally satisfied. I would attend to these jobs later. The fusible plug screwed into the crown plate was in satisfactory order. I now required to inspect the water side of a couple of the cross-tubes. Noticing a pair of 24 inch Stilson wrenches and a 3 lb hand hammer on top of a strong steel bench bolted to the wall, I asked Jake if I could borrow both tools. Puffing a cloud of reeking cigarette smoke into the air, he nodded.

Using the Stilsons, I quickly removed the nuts from two of the mud-lids which were carefully hammered in, and removed. Shining the torch through each of the mud-holes revealed both waterside surfaces of the cross-tubes to be in excellent order. **"Well young mon, ar't interested in me boiler?"** asked Jake while puffing on his tab. I had never previously purchased a steam boiler, but this VCT boiler subject to a successful hydraulic test appeared to be in good order. I also, had in mind one of my customers on Dandelion Colliery Piggeries, who had just a mere few weeks earlier, asked me to find him a replacement boiler. Therefore, I felt certain that he would purchase the Farrar VCT, and importantly, I knew that the Dandelion Pigger also had the necessary brass in his bank account.

I informed Jake that I was indeed most interested, but a sale depended on the results of a hydraulic test. After I explained the procedure, he told me to go ahead and carry out the hydro test. Jake and I walked out of the Boiler House and headed towards another building which he called 'Swill Cottage'. Upon noticing my obvious interest in the architectural details of the two storey, mullion-windowed building, which resembled an old farm worker's cottage, Jake explained, that the building was all that remained of the old farm that had been built on the site in the late seventeenth century.

When Whittle Brothers had originally purchased the site to search for coal, and with not finding any sign of the black stuff, subsequently instead mined the yellow clay, they demolished the farm house and attendant barns. How fascinating, I must have a good look round at a later date, I thought.

Inside the low-ceilinged building, were two rows of old beer barrels, above which was a rust-eaten iron pipe. This would have delivered steam into the barrels which would have been full of swill; the steam cooking the pig food. Located at the other end of the room, was a 3 feet diameter by 5 feet high, vertical, riveted steel, steam-jacketed pan with attendant steam and water pipe work still attached. I carried out a thorough visual inspection of the vessel and nodding my approval to Jake, I told him I was indeed, interested in purchasing it. Again, I explained the need for a satisfactory hydraulic test.

Jake consented to me carrying out both hydro tests in two days' time. But first, I would have to contact my valued customer, Jemmy Peck, the operator of one of the larger, more profitable Dandelion Colliery Piggeries. Meanwhile, poor old Jake was still coughing and retching as I bade him farewell.

Back in my tiny office in my beloved Boarshaw Boiler Works, I found myself tingling with excitement, for Jemmy Peck was very interested in viewing both the VCT boiler and also the steam-jacketed pan. Two days later, accompanied by my old chum, Harry Cunliffe, who had, with great enthusiasm volunteered to assist me to carry out the two hydraulic tests, I drove my van into the Old Golden Pits Piggeries, conveniently parking the vehicle close to the Boiler House door.

A pair of Vertical Steam-Jacketed Pans.

After I had cleaned, and fitted new asbestos joint rings, Harry and I refitted all of the mud-lids. Using my own rubber hosepipe, Harry started to fill the boiler through the man-lid. I asked my chum to keep his eyes open for leaks while I carried out the dismantling of the rusty steam and water pipe work on the steam-jacketed pan over in Swill Cottage. The boiler was full up to the man-lid within an hour. We then blanked off the boiler in readiness for the hydro test. Fortunately, Harry had reported no leakage, so we then mounted my hydraulic test pump. Up the ladder I went, and soon the boiler man-hole with new asbestos joint ring was refitted. Meanwhile, Harry had pushed the hosepipe into the opened-up safety valve stool on the boiler top. Another half hour ticked by as the incoming water filled the boiler's steam space.

With Harry enthusiastically working the plunger up and down on the hydro pump, I bled off the trapped air, then after closing the small bronze air cock, the pressure slowly commenced rising. I went around nipping up the nuts on the mud-lids we had refitted. I shone my torch into the firebox; no leaks were revealed. Upon attaining 100 pounds of pressure, I instructed Harry to keep pumping as I carried out a final visual inspection all over the boiler.

The hydraulic test went according to plan. Thus, I was extremely satisfied with our morning's hard work. The next job was the hydro-testing of the steam-jacketed pan. Harry pumped the vessel up to 50 pounds. While the vessel was under pressure, I sounded the plate-work both internally and externally with my hammer. The vessel proved to be in good condition. Again, I was most satisfied with the outcome.

While Harry was engaged packing my tools and hydraulic testing equipment into the Morris J4 van, I was sat with Jake in the office, carrying out some serious bartering which went on for over an hour. Eventually, Jake accepted my final offer of £150 for the VCT boiler and £30 for the steam-jacketed pan. The next morning bright and early, Jemmy Peck arrived to view both the boiler and the jacketed pan. He was clearly very satisfied and accepted my prices for both which included the dismantling, loading, cost of transport and delivery to Dandelion Colliery

Piggeries. I had requested Jake to keep out of the way, and he did so.

The same afternoon, Harry and myself, using the conveniently sited steel lifting beam situated high above the boiler and the 5-ton chain blocks suspended from a 'mule's ear' coupling, managed to lift the now drained-down boiler, which I estimated weighed around three tons. A considerable amount of heavy pulling and heaving with a small set of chain blocks and a pull-lift followed. Thus, the boiler was gently lowered down onto its side in front of the Boiler House doors. We then tackled the dismantling of the steam-jacketed pan, which weighed around one ton, and was therefore considerably easier to manoeuvre.

Early next morning, as previously arranged by myself, my good friend Alec, the Boarshaw Road coal merchant had managed to talk a friend of his, the operator/owner of a huge Scammell breakdown lorry to assist with loading the boiler and steam-jacketed pan and transporting them to Dandelion Colliery Piggeries. By later that day, the Vertical Cross-tube boiler was undergoing winching onto the back of the huge Scammell, diesel engined break-down lorry. Harry and myself, together with the assistance of two of Alec's coal men had built up a ramp of wooden sleepers and tin sheets. This made an easy job of loading the boiler, for the lorry's powerful winch effortlessly hauled the three ton boiler up the sloping timber ramp and onto the lorry's flat bed. Just as the loading was completed, my customer, Mr. Jemmy Peck returned to Old Golden Pits Piggeries with a wad of cash amounting to £475. This was my agreed price with him. Thus, I was over the moon.

Our next job involved jacking up the flat-bottomed steam-jacketed pan and placing several lengths of steel pipe underneath the vessel. Harry and myself and one of the coal men named Les, then man-handled the vessel by rolling it the full length of the Swill Cottage floor and out into the yard. Les then backed his Bedford lorry, end on to the vessel. We then set to constructing another timber and sheet metal loading ramp. On my instructions Harry fixed a set of pull-lifts onto a robust steel girder mounted onto the lorry's chassis at the back of the cab. Then both Harry and Les took turns at ratcheting the vessel up the timber slope until it was safely positioned on the load bed.

The job went excitingly well. As soon as Jake returned to Old Golden Pits Piggery I handed him the agreed sum of £180 in cash. We shook hands and he promised to buy me a couple of pints in my local pub, The Red Lion.

The loads were safely delivered to Jemmy Peck's Piggery at Dandelion Colliery, where they were quickly off-loaded by an Iron-Fairy crane, whose hire was paid for by Mr. Peck. Late that exciting, eventful and extremely profitable afternoon, I handed my coal merchant friend Alec a wad of cash and thanked him for his kindly assistance. He would then pay his friend the owner of the Scammell break-down lorry.

Jemmy and his two beefy sons carried out the full installation of both the Farrar VCT boiler and the steam-jacketed pan at their Piggery. As per my promise, I subsequently carried out the electric-arc welding-'buttering' of the spots of corrosion on the boiler uptake tube. Jemmy Peck was most satisfied, and consequently broadcast my name far and wide.

A fortnight later, Harry and myself were enjoying a drink in The Gardner's Arms when Percy Ramsgill delivered the sad news that Jake Belthorn's cancer had taken his life.

Chapter 7

<u>BESSIE ROGGERHAM'S BLUEBELL NOOK PIGGERY JOB.</u>

$\mathcal{O}n$ a cold, bright yet beautifully sunny early February afternoon in 1970, I found myself at Bessie Roggerham's Piggery in delightful, verdant Bluebell Nook betwixt Clayton and Ashton under Lyne. Earlier that morning, I had received a phone call in my office at Boarshaw Boiler Works. The voice at the other end was that of Bessie Roggerham, the owner of a large, long-established Piggery. She informed me that the Piggeries' Cochran Vertical Boiler was broken down due to a serious leakage of water causing dire problems. She impressed upon me that without steam her pig-men could not boil the huge quantities of swill needed for the 2,000 hungry pigs she kept. The woman was obviously deeply worried, and despite me being busy with on-going work, I assured her that I would drive over and inspect the boiler right away. She then gave me the address of the Piggery and the directions.

I telephoned my good friend Harry seeking his assistance, who said he would be most willing to help me. After I loaded a couple of tool boxes into the Morris J4 van, I then picked up Harry from near the Drill Hall on Manchester New Road. We drove past the gigantic Accrington brick – Warwick Mill, then took the road passing Middleton railway station, thence onwards to Middleton Junction and Chadderton until we reached the A62, the main Manchester to Oldham road; from here we slowly trundled through two very pleasant farming hamlets to eventually descend into the Medlock Valley. Nestled in this tranquil and beautiful vale, there were a number of pig farms, whose swill boiling activities delivered volcanic eruptions of huge clouds of steam that delivered the most peculiar aroma to pervade the whole surrounding district.

Following Bessie Roggerham's description of Bluebell Nook Piggery and her directions, which I had jotted down on a pad, now clamped onto the metal dashboard by a strong magnet, while turning a bend in the narrow, twisting lane, I noticed up ahead a grouping of historic buildings, which I guessed, had once been a water-powered corn mill. Located on the gable end of a semi-derelict barn-type building was the crumbling remains of a large cast-iron and timber water wheel. There was a bright blue painted sign on the side of an old farm cart proclaiming: Elizabeth Roggerham & Co. Bluebell Nook Piggery.

At the entrance to a cobbled yard leading into the interior of the collection of buildings, I pulled up the Morris J4 van, Harry and I climbed out. Casting my eyes over the architectural details of the buildings which were built of old, hand-made bricks, with large sandstone quoins at the corners, I guessed that their origins dated back to the late 17th century. Echoing from all around the historic buildings was an almighty, terrible din of squealing, grunting and shrieking pigs. My guess was that the porkers, due to not having being fed with swill, were therefore, extremely hungry and becoming troublesome.

Looking up the yard, I noticed a building on the left with a squat, hexagonal tapering brick chimney poking out of the roof. This clearly was the Boiler House. We jumped back into the van and I drove it up the yard and parked it close to a large, round-topped set of doors. Just as we were climbing out of the van, the confines of the yard were filled with the roaring sound of a diesel engine lorry which made us both stop and look behind us.

A large Bedford TK lorry, carrying a great number of metal dust bins was slowly advancing down the greasy surface of the cobbles. We leaned up against the Boiler House doors to allow the lorry to safely pass us. I noticed that the lidless metal bins on the back of the lorry were brim-full of waste food; a revolting mess of what looked like canteen dinners. The smell of the rotting food then engulfed us. **"Oh, my word, that stinks terrible."** Said Harry wrinkling up his nose.

Harry and I were already attired in boiler suits, steel toe-capped boots and tweedy caps, so carrying my battery-powered torch in one hand, and my small inspection hammer in the other, we entered through one of the doors and into the interior of the Boiler House, which was dominated by a huge Cochran Vertical No. 17, steam boiler. I noticed that it was fired by a James Proctor mechanical stoker, which conveyed coal from a large brick-built bunker. There was evidence of feed-water leakage from the smoke box doors, so grabbing a small ladder propped up against the wall, I climbed up and opened up both sets of catches.

A 5 feet diameter Cochran Vertical Multi-tubular Boiler, with a maximum working pressure of 250 p.s.i.

Harry and I pulled open wide both doors. It was then plain to see the water leakage was coming from several smoke tubes at the top of the left-hand tube bank. I ran a tape measure down the length of a tube and noted the length and diameter in my notebook. I estimated that there were probably eight leaking tubes. What I had thought was a relatively small job had now turned out to be much bigger. I would therefore, have to find a telephone and ring my boiler tube supply specialist firm down in Trafford Park.

Suddenly, in walked a pleasant looking, stoutish man, aged around 30 sporting a large unruly mop of blonde hair. He was attired in a pair of heavily stained, brown corduroy trousers and an ill-fitting pullover that appeared to be two sizes too small. On his feet were a pair of wellington boots, the tops turned down into which his trouser bottoms were thrust.

"Hiya lads. Bessie an' me are really glad you've come, for we're knackered without steam. Our pigs are dreadfully hungry, and we've already lost a few. Bloody famished pigs soon turn into cannibals. The pigs kill and eat their own kind." Said the man sadly. **"By the way, I'm Tim Snape, the Head Pigger. My boss is Bessie Roggerham. You'll meet her soon. She's one heck of a hard woman, but fair. She's been good to me. She's with the other lads feeding the bloody, murderous, squealing pigs with highly expensive corn."**

I quickly introduced myself and my chum Harry, and informed Tim that the boiler required a number of tubes replacing which would require us having to drive down to Trafford Park to collect, and also to bring my Bedford lorry and other boiler repair tools and equipment onto the site. Nevertheless, I assured him that we would commence the boiler repair that very day. He then invited us to follow him to the office where there was a telephone.

"Coom on lads. This way." Harry and I followed the cheerful, stout, blonde-haired Pigger out of the Boiler House and up the cobbled yard. The all-pervading reek of pig muck and pig swill made us cough and retch. ***"Eeh lads, that stink is nowt. Shiftin' the half-eaten carcases of pigs with their guts hanging out, reeks far worse."*** Said Tim smilingly. He led us into a wide loading bay. The Bedford lorry we had earlier seen, had now backed in and a number of long-haired, hippy-looking young chaps were busy unloading the bins of the revolting-looking waste food.

Tim motioned us to follow him up a short flight of extremely slippery concrete steps leading onto the floor above. The surface of the cracked concrete floor was even more greasy, due to the trod-in waste food and offal. ***"Take it steady lads, we don't want you fallin' into the swill cooking pan."*** Chortled our new mate Tim, while pointing to a huge, riveted, steel, horizontal steam jacketed-pan set into the floor with the top jutting up about six inches above floor level.

Looking down into the depths of the pan, which I estimated was about six feet deep, and twelve feet long, I shuddered with disgust. Hell's Teeth! – when that large steam-heated vessel was chock-full of boiling pig swill the stink in this room would be truly nauseating, I thought. I also noted that there was no encircling protective fence or barrier to stop anyone falling into the pan. What a terrible death to be boiled alive. Again, I shuddered.

Tim was still merrily chatting away. ***"This 'ere pan's daily output of boil't swill, and together with the valuable by-product of fat, is the heart of Bessie's Piggery business."***

He then gave us a mini lecture informing us that most waste food has a high fat content. When boiled into pig swill, the fat floats on the surface. This fat, known as 'technical fat' is valuable and is purchased from the Piggeries by a specialist fat refining company, who, after refining the fat, sell it by the lorry tanker load to a wide range of firms, such as cosmetic manufacturers, who use the fat as a base in products such as women's lipstick, eye-liner and body oils.

"Without steam, the whole job is at a standstill, and we're earning no brass. Reet, lads, foller me, an' I'll tek you to meet me gaffer, Bessie Roggerham."

Tim then commenced whistling a merry tune as he led Harry and I through the extensive Piggery buildings which loudly echoed with the squeals and blood-curdling shrieks of the starving pigs as they raced around their sties in what had become death carousels. The pig sties ran either side of narrow ginnels, flanked by walls of about 3 feet, 6 inches high. Harry and I agreed that the pervading atmosphere, and the nauseating smells causing us to continuously cough and retch, and the loud snorting, grunting, honking and screaming frenzies of the dying pigs were truly revolting and horrific. Looking ahead into the semi-gloom, I suddenly saw a very large woman. Tim shouted a greeting. ***"Hiya Bessie, I've got Alan the boiler man and his mate Harry here with me. Alan urgently needs to use the telephone."***

"Good morning gentlemen, which one of you is Mr. McEwen?" questioned the posh-sounding, commanding voice.

Bessie Roggerham was of huge stature, a female giant; an Amazon of a woman. She must have stood 6 feet 5 inches in her heavy, thick-soled gum-boots. Perched on her head, she wore a sort of cowboy Stetson. Her long thick hair was the colour of burnished copper and hung down to her massive, broad shoulders. Her eyes were a beautiful emerald green. The woman's fleshy cheeks were flushed, and I noticed her strong, angular jaw. She had large masculine hands with nails bitten to the quick. Bessie wore a blue boiler suit and a large donkey jacket. Moving forward, and feeling rather mesmerised at the sight of the female giant, who dwarfed me in height, I introduced myself and Harry.

I then informed her that the boiler required several new tubes which, to save time, we would collect from the merchant in Trafford Park. Thus, after ordering the tubes via the telephone, we would then go back to Boarshaw Boiler Works to collect my Bedford lorry, the necessary tools and equipment prior to travelling to Trafford Park. Therefore, it would be later in the day when we returned to commence the boiler repairs.

Bessie then went into some detail of explaining their dire situation and the urgency to expedite the Cochran Vertical boiler repairs. I gave her my promise that upon our return we would get stuck in to the work, and if necessary work right through the night to replace the leaking smoke tubes and return the boiler to steam. I asked Bessie if she would telephone the Boiler Inspector to ask him if he would visit during the evening. With the sweep of her large hands, she informed me that like the other local Piggers, she didn't bother with specific boiler insurance as her business insurance policy covered the boiler anyway. All four of us shook hands, and then Harry and I made our way back to the Morris J4 van.

Due to an ever-increasing influx of larger, heavier boiler and fabrication jobs, my small engineering business urgently required the acquisition of a small lorry. For over the past six months, I had been earning some good money. My building society account, that I used as a bank, held over £1,000: truly a King's ransom. It was this capital that had enabled me to purchase for the sum of £120, from my friend Alec the Coal man, a low mileage, Bedford TJ drop-side lorry, which was in excellent mechanical condition, with brilliant navy-blue paintwork. I commissioned a local sign writer to paint signs onto the cab doors and drop sides:

H.A. McEWEN (BOILER REPAIRS),
BOILER REPAIR SPECIALISTS,
BOARSHAW BOILER WORKS,
MIDDLETON, LANCASHIRE.
TEL: MID 6008

The Bedford was powered by a Vauxhall Cresta 2.2 litre petrol engine, which unfortunately delivered a rather poor 18 m.p.g. Rather costly to run but a most reliable, handy vehicle. I built and fitted to the rear of the cab, a pipe and tube carrying rack. While attending an auction at a closed-down boiler works in Warrington, for £40, I purchased a Broom & Wade diesel compressor, together with rubber air lines, an Atlas Copco 1 inch square drive impact-wrench, which was just the ticket for tube expanding, together with a wooden box of around 50 tube expanders, mandrels, spare rollers for £18. This was money well spent.

Alec allowed me to store the Broom & Wade compressor under a tarpaulin sheet in a quiet corner of the Coal yard. When required, it could be simply lifted with one of his loading shovels directly onto the bed of my lorry.

I used the Bedford lorry to collect twelve, 2 inch boiler tubes from Trafford Park. Upon returning to Boarshaw Boiler Works where I had left Harry to gather together a list of tools and equipment, including oxy-propane cylinders and burning gear, we then loaded up the lorry, including getting Alec to load the Broom & Wade compressor and airlines.

It was past 6 o'clock in the evening, and dark, when I drove the Bedford lorry into the yard of Bluebell Nook Piggery, and parked it up in front of the Boiler House doors. The whole Piggery still echoed with the sound of pigs screaming and squealing. Jumping down from the cab, both Harry and I quickly unloaded the oxy-propane burning gear, a large metal ex-British army Mills bomb tin full of tube expanders and mandrels, another box containing hand hammers and a 7 lb short-shafted sledge-hammer, a crocodile gob and a sundry collection of other tools and equipment. We also unloaded two pairs of 5 feet - 6 inch tall stillages I had previously fabricated from old steam pipe; together with a quantity of second-hand scaffolding boards.

James Proctor Ltd. Mechanical Stokers fitted to a Lancashire boiler.

We set up one pair of stillages at either side of the Proctor stoker and placed the scaffolding boards on top. Using the wooden ladder that I lashed with rope onto the right-hand stillage, I clambered up the rungs to test our 'Heath Robinson' working platform. I swung both of the heavy smoke-box doors as far back as possible, and Harry tied them back with rope. The front tube plate was relatively free from soot and dust, but within the smoke-box base was about 6 inches of fly-ash. I instructed Harry to clear it away. I then clambered down and rigged up a 240-volt electric hand lamp. I clambered back up the ladder and clipped the hand lamp onto a short length of pipe which I inserted into one of the smoke tubes.

Our next stage involved tackling the heavy, hazardous job, of dismantling and removing the large, curved, firebrick-lined combustion-chamber door. Thus, as Harry and I squeezed through the tight space between the Boiler House wall and the boiler shell of the huge Cochran Vertical Boiler, carrying another 240-volt hand lamp, the loose asbestos lagging covering the boiler fell upon us in great lumps causing the dust to fill our noses and throats.

Eventually, with much coughing and retching, I wormed my way forward to the rear of the boiler and upon looking up, whilst shielding my eyes to avert the falling dust, noticed a huge set of chain blocks suspended from a chain wrapped round a steel girder set into the roof of the building. The chain block hook was already attached via a sling shackled through both lifting lugs on the large curved combustion-chamber door. I clipped the hand lamp onto a handy metal bracket protruding out of the wall. I climbed the iron ladder bracketed to the rear off-side of the boiler shell to access the large plate dogs securing the door. These were removed.

We then commenced to operate the chain block to take the weight of the heavy, firebrick-lined combustion-chamber door. Using a long pinch bar to pull it away from the boiler, we slowly lowered it to the floor. We now had unrestricted access to the rear tube plate for the removal of the eight leaking tubes. Harry handed me the second set of stillages and scaffold boards and then joined me to erect the working platform. Thus, we were now ready to proceed with the removal of the tubes.

However, the day had gone by in a flash and both Harry and myself were suffering from severe hunger pangs, but unfortunately, we hadn't brought any sustenance with us. Leaving Harry to generally set up the tools and equipment, I made my way to where I thought I could perhaps find Tim. As I walked further up the dark yard, using my torch to light the way, the sound of my boots on the cobbles caused hundreds of squealing rats to flee. What horrible creatures I thought, my right hand grasping the shaft of my hand hammer stuck through my leather belt. Then from out of a doorway up ahead, Tim suddenly appeared.

"Hello again Alan, I am glad you've returned. Did you pick up the boiler tubes and the other tackle? I thought I heard a wagon drive in a bit back." I updated him, and also mentioned how hungry both Harry and myself were. Smiling, Tim said how he would get Bessie's housekeeper, Dolly, to make a plate of ham butties and pots of tea. However, I made it clear that neither Harry nor myself would be able to stomach bacon or ham. Tim laughed, and said he would instead get Dolly to rustle up some cheese butties. He advised me to return to the Boiler House and await the grub. A short while later, Tim returned carrying a platter of thickly cut, bread and cheese sandwiches, slices of fruit cake and large pots of tea. We thanked him for the tasty-looking repast, and after washing our dirt-stained, greasy hands under the Boiler House cold tap, we then got stuck in wolfing down the food.

While Harry and I were filling our bellies, Tim chatted to us good-heartedly. He mentioned that apart from his job as the Foreman Pigger, he was also Chief Mechanic, Electrician and general handyman and also responsible for his team of lorry drivers who daily collected waste food from houses, council estates, restaurants and shops in the local towns. Three times a week they also picked up several tons of waste cake from a large bakery. The Piggery operated a fleet of five Bedford TK, drop-side lorries and a Bedford CA van to collect the waste food and deliver it into Bluebell Nook Piggery.

The Cochran Vertical boiler was around 16 feet tall with a diameter, including the thick asbestos insulation amounting to over 7 feet. The boiler worked at 50 lbs and delivered steam directly into an asbestos lagged pipeline that ran throughout the Piggeries. A pipeline carrying condensate came back from the far reaches of the Piggery and discharged into a large hot-well tank.

I clambered up onto the scaffolding at the front of the smoke-box, and got stuck in with relish, using my burning torch, cutting V-shaped notches into the ends of the 10 condemned smoke tubes. After a careful extra scrutiny, I had decided to cut out two more tubes. For many, this burning process would be a seriously, risky, tricky job but for over seven years I had removed literally thousands of tubes by this process. Thus, it was all in a day's work! This process resulted in loosening the tubes. It only took me about 40 minutes to complete V-notching the tube-ends on both the smoke-box tube plate and the combustion-chamber tube plate.

I then commenced driving out the now loosened tubes, using my 7 lb, short-shafted sledgehammer with Harry holding the 'Crocodile-gob' steel knocking-out bar. Despite being thickly coated with more than a quarter inch of loose scale, it took a mere three or four belts with my sledgehammer. Thus, one after the other the tubes were driven out of the combustion-chamber tube plate. We then moved around to the scaffold boards at the front of the boiler.

Placing my leather-gloved hands around the first tube, I wiggled it to remove the loose scale, then yanked it out of the smoke-box tube plate. I gave it a good dose of eyes-on inspection which revealed the small localised corrosion that had caused leakage. I handed the tube to Harry who placed it on the back of the Bedford lorry. Thus, one after the other we quickly managed to remove the ten condemned smoke tubes, all of which were stored on the back of the Bedford. I instructed Harry to clean out the tube holes and wire brush the tube plates. My V-notch burning had gone well and fortunately, there were no troublesome nicks in the tube plate. We had earlier connected up the rubber compressed air pipeline onto the air-outlet valve on the large diesel Broom & Wade compressor which was chained down onto the back of my Bedford lorry. Close by were two, 10-gallon drums of red plant diesel fuel for the compressor. I attached my Consolidated Pneumatic tube expanding machine to the end of the rubber airline. We were now ready for re-tubing the boiler.

We had previously placed 10 out of the 12 brand new boiler tubes taken from the back of the lorry onto the floor at the front of the boiler. Thus, I got back onto the scaffold boards in front of the smoke-box. Harry handed up the first boiler tube, which I inserted into the lower of the tube holes in the tube plate. I then pushed the tube well into the boiler. This process was followed by fitting tubes into the other four bottom tube holes. I then instructed Harry to clamber up onto the combustion-chamber scaffold boards, and by using a two feet length of ¾ inch pipe inserted into a bottom row tube hole, allowed the tube to be guided into the rear tube plate. As soon as Harry shouted that he had the smoke tube on the end of his pipe, I gave the tube a push and then upon it engaging into the tube hole, drove it home with a couple of hammer blows. This process went on until all ten tubes were fitted.

I then instructed Harry to insert tube clamps into the tube ends at the combustion-chamber tube plate. All ten tubes were now firmly locked into place. I requested Harry to join me on the smoke-box scaffold boards. Upon him clambering on, I passed to him a, two-inch, 5 roller tube expander and a small oil can. I instructed him to squirt a drop of oil on to the tube expander, then place it into the tube hole, and then engage the mandrel. I jumped down off the scaffold boards and walked through the Boiler House open doors and over to the Bedford lorry. I heaved myself onto the platform, and then fired up the powerful Broom & Wade compressor. I then re-joined Harry.

I showed him the ropes by placing the C.P. tube-expanding machine onto the square end of the mandrel and by opening up the twist grip the machine roared into life. The rapidly revolving tube-expander quickly expanded the tube. With a nod to Harry I then passed him the tube expanding machine, and I indicated to him to use the machine in reverse to remove the tube-expander. He soon grasped the procedure, and then carried on tightening up the ten tube ends on the smoke-box tube plate. Whilst Harry was thus engaged, I carried out a survey around the mud-lids on the boiler shell, and then mounting the fixed iron ladder, clambered up to the domed boiler top, where, after a struggle, and covered head to toe in asbestos dust, with much coughing and retching, I managed to remove the safety valve and fit a blank flange with an air cock. Looking at the boiler man-lid, which being a Cochran Vertical could be difficult to remove and replace, I checked to see if the joint was tight. Indeed, it was, so I thankfully decided to leave the man-lid undisturbed.

The boiler was fitted with a new looking electric Worthington Simpson feed water pump, therefore, after opening the feed water valve on the adjacent large hot well tank, I threw the electric switch causing the pump to commence filling the boiler. Dependant on the cold-water supply feeding the hot well tank expediently, thereby to fill and bleed air out of the boiler, hopefully should only take an hour. Until then, it was like what soldiers and policemen regularly endure: a waiting game!

While the boiler was filling, I was repeatedly checking for leaks on mud-lids and also on both front and rear tube plates. All appeared to be tight. Due to the Boiler House now being quiet, we could still hear the sound of much squealing and shrieking emanating from the pig sties. I was wondering how Bessie and Tim were coping, when in through the door walked the man himself. He asked how the repair had gone and I filled him in. Five minutes later, I could hear air being blown out of the air cock mounted high on the top of the boiler. Up the iron ladder I went. I stayed on the boiler top until a jet of pressurised water hit the ceiling showering me with muck and asbestos. I shouted down instructing Harry to switch off the water feed pump and down the ladder I went.

Firstly, I inspected the combustion-chamber tube plate. Thankfully it was dry. I shouted to Harry to take a look at the smoke-box tube plate. He reported there was a tiny drip on one tube end. I instructed him give it a nip with the tube expander and ratchet. Two minutes later he confirmed it was dry. I jumped down and walking round to the feed water pump, brought the pressure up to 25 p.s.i. Using Harry on the front end, and me at the rear again we looked for leakage. It was dead tight. I asked Harry if he would take up the pressure to 40 p.s.i. while I remained at the rear of the boiler. This test also proved the boiler to be tight. We then carried out the final test of 75 p.s.i., which proved satisfactory. We had therefore, achieved our goal.

After the heavy job of remounting the combustion-chamber door plate and securing the two smoke-box doors, the blank flange on the boiler top was carefully removed and the safety valve refitted. Using the blowdown valve, I drained the boiler down until it showed half way down the water gauge glasses. Tim reappeared, and upon hearing from me that all was well, was clearly over the moon. He immediately started preparing the James Proctor Stoker, and after throwing in a bundle of diesel-soaked rags and wooden sticks into the firebox, he applied a match.

"Alan and Harry, you've returned the good health of our boiler. While my mechanic, Charlie, gets steam up, how do you fancy takin' a gander around our Bluebell Nook Piggery, which we're proud to boast is one of the biggest in Lancashire?"

I then noticed the mechanic Charlie, who had quietly walked into the Boiler House. I looked at Harry who gave me the thumbs up, and then informed Tim that after we had cleared and loaded up all our tools and tackle we would be delighted to accompany him around the Piggeries. With gusto, we both got stuck in to dismantling the scaffold platforms and to clearing away all tools and equipment which were carefully loaded onto the Bedford lorry.

Later, it was indeed, a most fascinating experience to walk down the many criss-crossing ginnels that accessed the numerous pig sties. However, the terribly offensive sights, sounds and smells were quite awesome. It was horrible to behold the sight of dozens, of hungry, yet strong pigs, - chasing weaker, younger pigs in circles around the confines of the sty. The poor terrified weak pig, foaming at its mouth, would drop onto the sawdust strewn floor totally exhausted.

Then the strong porkers with open jaws revealing rows of gleaming spear-like teeth, would launch themselves onto the defenceless, almost dead, fallen pig. The ripping open of the poor unfortunate pig's belly would then commence. Spurting blood would cover the snouts of the attacking pigs. More pigs scenting the reek of the dead pig's stomach contents would then join in the grisly, blood-soaked chomping and flesh-ripping melee. The cannibalistic porkers' powerful jaws and sharp teeth would quickly bite off large chunks of the pig's body; guts, snout, eyes – until there was nothing left other than the blood-soaked sawdust on the floor.

I had previously witnessed this spine-chilling, horrific spectacle several times when working on other Piggeries. The sight of squealing, shrieking and hungry, terrified pigs is disgusting to behold.

The first time I witnessed cannibalistic pigs on the rampage, I was almost transfixed with the shock and horror as I beheld the awful scene; the vomit rising up in my throat. My head was filled with the death squeals of scores of dying pigs as their bodies were being ruthlessly hacked into food for the packs of large, extremely strong, dangerous pigs. However, despite my body shaking with nerves, and bottling the urge to jump into my van and scape from the Piggery, instead, my professional training kicked in. I just rummaged into the old Austin Devon for one of the bottles of cold Horniman's tea, my dear mother had that morning brewed for me. I took a long swig which made me feel much calmer. Hornimans tea was far better than ale; I had always enjoyed drinking the brew. However, Barley wine tasted better!

Upon returning to the Boiler House, Charlie had managed to successfully steam the boiler. Tim then invited both myself and Harry up to Bessie's large house on the hill at the top of the yard. Here we were cordially thanked by Bessie Roggerham, who urged me to send my bill in the morning which she assured would be immediately settled.

So, ended our adventures at Bessie Roggerham's Bluebell Nook Piggeries.

The Stoker stands at the front of a coal-fired Cochran Vertical Boiler.

An Old Boiler Maker's Scrapbook of Old Boilers.

Haystack or Balloon Boilers (For details of early steam boilers see my book: HISTORIC STEAM BOILER EXPLOSIONS: ISBN 978-0-9532725-2-5. Sledgehammer Engineering Press Ltd. 2009.)

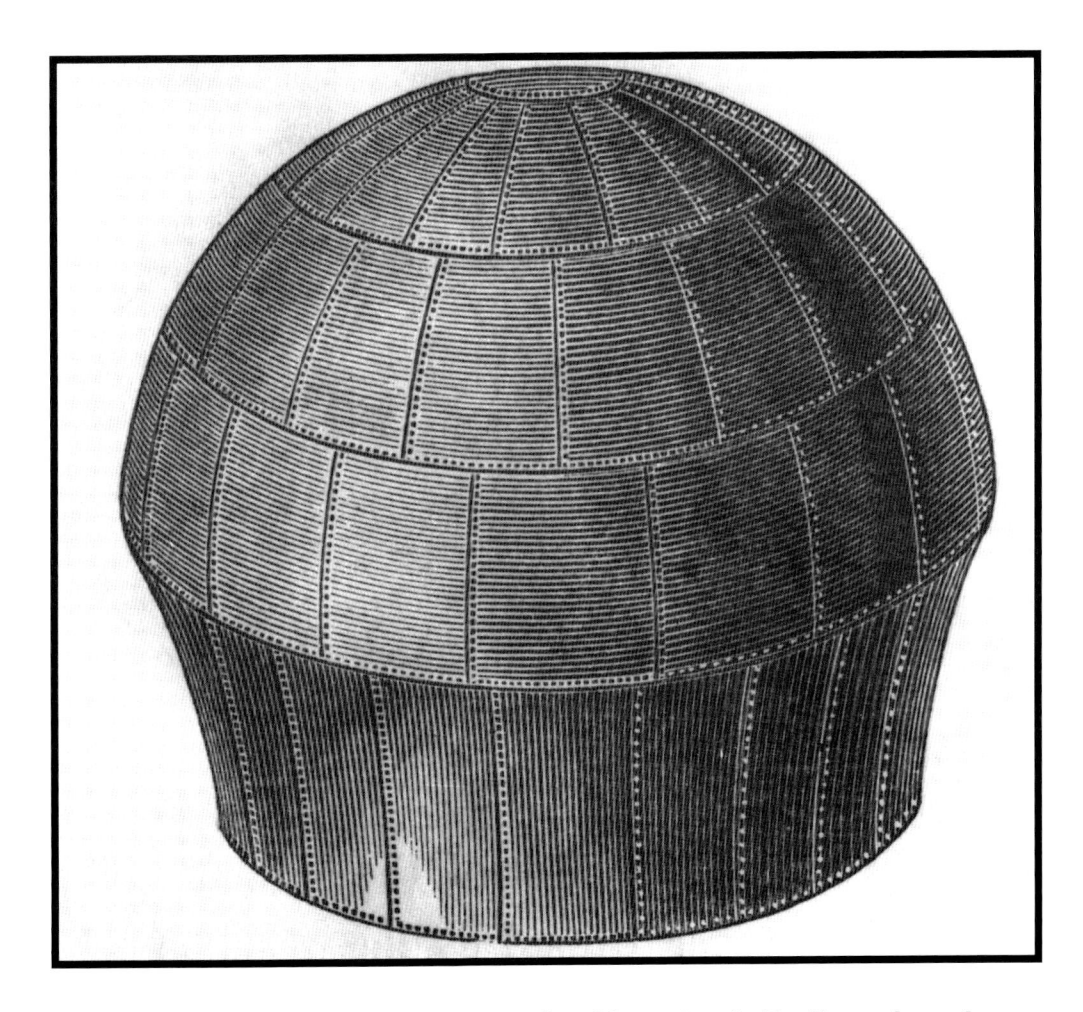

A 19th century engraving of a Haystack Boiler showing details of the small, narrow wrought-iron plates and lap-riveted seams used in the construction. Many of these Haystack Boilers were built in diameters up to 20 feet. They were commonplace in the Staffordshire iron-working industries, where they were known as 'Balloon Boilers'.

A Haystack Boiler photographed by the British Engine & Boiler Co. in 1938 in a field near Ashton-under-Lyne. Notice the large irregular-shaped riveted patch beneath the man-hole, and sundry other poor-quality repairs in the foundation area.

In 1947 this Haystack or Balloon Boiler was discovered in a field at Denby, Derbyshire. The boiler is considered to have provided steam at Bassett Colliery, Denby. It is now displayed at the Science Museum, London.

©Photograph courtesy of D.E. Potter and Eric Potter.

The Denby Haystack Boiler loaded onto a lorry trailer, bound for the Science Museum, London.

©Photograph courtesy of D.E. Potter and Eric Potter.

Bassett Colliery, Denby, Derbyshire circa 1885. The Haystack Boiler providing steam to Newcomen-type 'atmospheric' engines for winding operations.

A Haystack Boiler built circa 1770 at Cheddleton Flint Mill Museum, nr. Leek, Staffordshire.

A Waggon Boiler of 'improved' design and of single-lap riveted construction, the upper section end-plates of convex shape; the lower strakes flattened. Otley Mills, West Yorkshire.

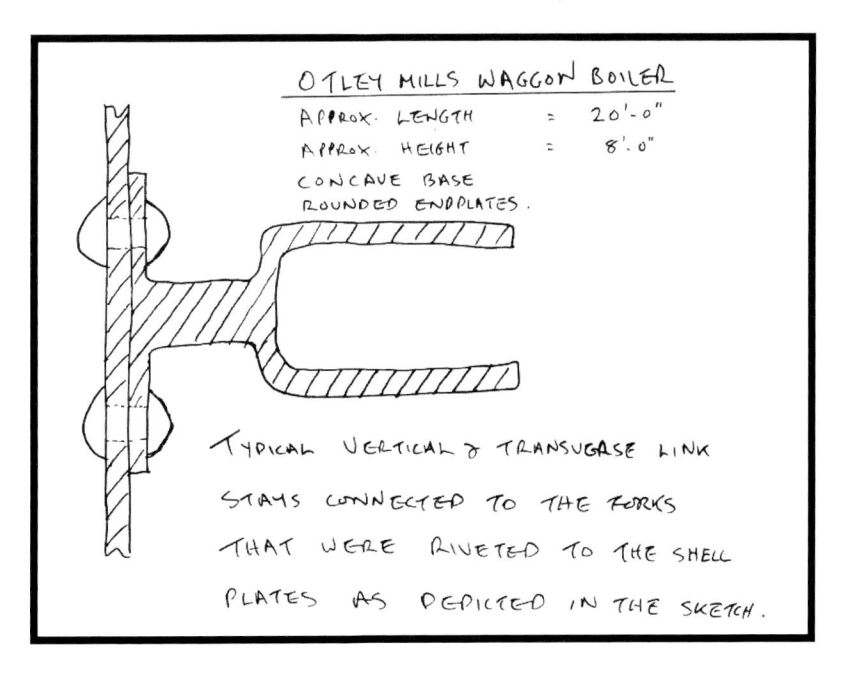

OTLEY MILLS WAGGON BOILER
APPROX. LENGTH = 20'-0"
APPROX. HEIGHT = 8'-0"
CONCAVE BASE
ROUNDED ENDPLATES.

TYPICAL VERTICAL & TRANSVERSE LINK STAYS CONNECTED TO THE FORKS THAT WERE RIVETED TO THE SHELL PLATES AS DEPICTED IN THE SKETCH.

Evidently, this Waggon Boiler provided steam for Otley Mill's Beam Engine. In 1985, when I discovered the boiler, it had for a great many years been used as a sewage storage tank. However, I was so intrigued to evaluate the internal construction that I clambered inside. In my notebook I made the above rough sketch of the Vertical and Transverse link-stays.

A Waggon Boiler of basic configuration and built of riveted wrought-iron plates being used as a water storage tank.
©Photograph courtesy of D.E. Potter.

The remains of a Waggon Boiler discovered on the Hopton Incline, Cromford & High Peak Railway, Derbyshire.
Photograph 24 July 1994.

This is the largest collection of Egg-Ended Boilers in the U.K.
Beamish Museum.

Hemispherical-Ended or 'Egg-Ended' Horizontal Steam Boiler
built of riveted wrought-iron plates, circa early 19th century.
Used as a water storage tank on a farm.

Cornish Boiler built circa 1870, which was fired off wood waste to generate steam for heating the premises of W.J. & R. Thompson (Wood Turners) Ltd of Sutton-in-Craven, North Yorkshire. Photograph May 2003.

I recollect carrying out a welding repair on the furnace of this old Cornish Boiler on Christmas Day, 1972.

Cornish Boiler built by Gimson & Co, Leicester, 1902, with Hodgkinson under-feed stoker. All Saints Brewery, Stamford, Lincolnshire.
©Photograph courtesy of Geoff Hayes.

Dish-Ended Cornish Boiler built 1925 by John Thompson & Co. Ltd. of Wolverhampton. New Star Brickworks, Leicester.
©Photograph courtesy of Geoff Hayes.

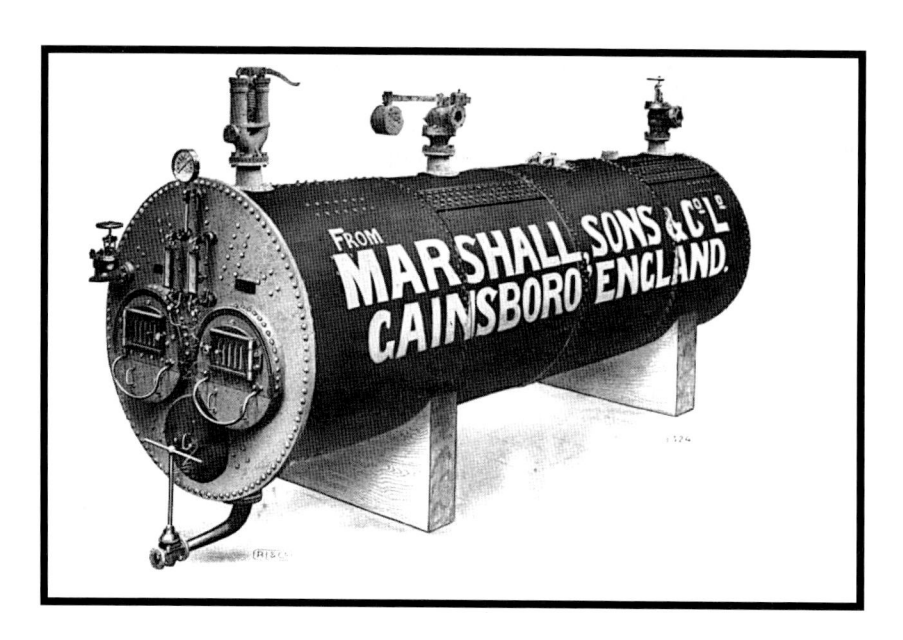

A Lancashire Boiler built by Marshall, Sons & Co. Ltd. of Gainsborough. Note the four strakes that make up the boiler shell and also the heavily riveted butt-straps. On top of the shell is a Ramsbottom Safety Valve; in the centre a high-steam Low Water Compound Valve and at the rear an Angle-Pattern Main Steam Valve.

A bank of 12 Lancashire Boilers, 30 feet long x 9 feet diameter in a South Yorkshire Colliery long since demolished.

The author standing alongside a pair of Vertical Cross-Tube Boilers that have been sectioned for display purposes. Summerlee Museum, Lanarkshire.

This is the picture for Boiler Making buffs!

The boiler that the author is leaning on is a sectioned Cochran Vertical, one of the most famous of Vertical Boilers in the world and originally constructed at Annan in Dumfriesshire. The other boiler is a typical Vertical Cross-tube Boiler, once built by numerous Boiler Making firms all over Scotland and the rest of the United Kingdom.

The Cochran Vertical Boilers were extremely efficient at steam raising and size for size could generally out-perform most other types of Vertical Cross-tube Boilers. Both types were used in virtually every industry ranging from 'donkey' boilers for marine use, laundries, dairies and multifarious industries including pig farms. The author has repaired scores of both types of boilers on sites ranging from Piggeries, to Chemical Works all over Britain.

OTHER QUALITY BOOKS FROM SLEDGEHAMMER PRESS LTD.

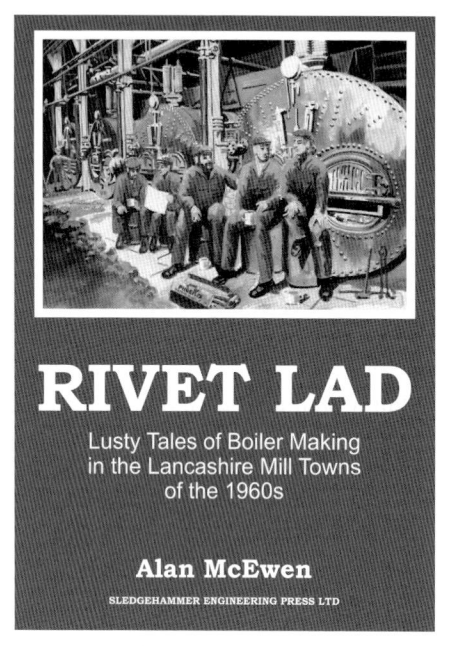

RIVET LAD £30.00 each including p&p
OTHERS £16.00 each including p&p
www.sledgehammerengineeringpress.co.uk

H.A. McEwen (Boiler Repairs) Ltd. Began Trading On The 4th Of August 1968

AUGUST 2018
H.A McEwen (Boiler Repairs) Ltd
Industrial & Heritage Boiler Engineers

Monday	Tuesday	Wednesday	Thursday	Friday	Saturday	Sunday
		1	2	3	4	5
6	7	8	9	10	11	12
13	14	15	16	17	18	19
20	21	22	23	24	25	26
27	28	29	30	31		

The Founder, Alan McEwen.

50 Years

EST. 1968

H.A McEwen Ltd
Farling Top
Cowling, Keighley,
West Yorkshire BD22 0NW
Phone:
01535 634-674

Fax:
01535 636-802
E-mail
enquiries@mcbo.co.uk
Website
www.mcbo.co.uk

SLEDGEHAMMER ENGINEERING PRESS LTD NEW BOOK

COMING SOON!

THE STEAM CRANE MAKERS OF YORKSHIRE

by

ALAN McEWEN